CANADIAN HEAVY EQUIPMENT TECHNICIAN

Certificate of Qualification
Test Preparation

Robert Huzij

Centennial College Press

Library and Archives Canada Cataloguing in Publication

Huzij, Robert, author
    Canadian heavy equipment technician : certificate of qualification test preparation / Robert Huzij.

ISBN 978-0-919852-73-0 (softcover)

    1. Machinery--Canada--Maintenance and repair--Examinations, questions, etc. 2. Industrial equipment--Canada--Maintenance and repair--Examinations, questions, etc. I. Title.

TJ148.H85 2017   621.8'16076   C2017-900926-5

Copyright ©2017 Rober Huzij

Centennial College Press
951 Carlaw Avenue
Toronto, Ontario
M4K 3M2

All right reserved. This publication is protected by copyright and permission should be obtained from the publisher prior to any reproduction, storage in a retrieval system, or transmission in any form by any means—electronic, mechanical, or otherwise.

Book design and typesetting Laura Boyle

Printed in Canada by Webcom

Contents

Chapter 1
    Introduction to the First Edition
        ➤ The Objective of This Book   1
        ➤ Who Qualifies to Write the Red Seal Exam?   1
        ➤ Scope of the Heavy Duty Equipment Red Seal Exam   3
        ➤ Understanding the Basics of Trade Knowledge   5
        ➤ How to Use the Practice Exams   8

Chapter 2
    Using the Practice Exams and Exam Profiles   9
        ➤ Example of a Red Seal Question   9
        ➤ Taxonomy Levels for Exam Questions   11
        ➤ Study Planning and Tips   12
        ➤ Using the Exam Profile to Weight Your Study Time   13
        ➤ Red Seal Exam Breakdown   14
        ➤ Acknowledgements   16

Chapter 3
    Practice Exam 1   17

Chapter 4
> Practice Exam 2   41

Chapter 5
> Answers and Explanations for Practice Test 1   65

Chapter 6
> Answers and Explanations for Practice Test 2   83

Introduction ➤ 1

The Objective of This Book

The intent of this book is to provide apprentice or trades qualifiers with a comprehensive guide to prepare to challenge the Heavy Duty Equipment Red Seal exam successfully. Having taught pre-exam courses over the past 25 years to hundreds of apprentices and trade qualifiers, I have come to realize that challengers have a difficult time in assessing the trade knowledge that is required to successfully challenge the Red Seal exam. A related problem is that challengers have difficulty self-assessing their mastery of the trade knowledge needed to pass the exam. In many work environments of large organizations, apprentices are not exposed to all areas of the trade frequently enough to develop expertise in each area of the trade. Therefore, the purpose of this book is to guide Red Seal challengers through a process that I have developed and used over the past 25 years to help apprentices and trade challengers succeed at the task of identifying what areas of the trade they need to study to successfully challenge the Red Seal exam.

The exam questions that have been developed for this book have been written in a manner that challenges the knowledge and skills developed by individuals who have worked in all 8 reportable areas of this trade in diagnosing and repairing equipment. In many cases the challenger will need to spend time relearning the basic principles of specific content in order to understand the questions used in the two practice exams in this book. Suggested reference material is identified later in this chapter. In most cases, the challenger will need to spend a considerable amount of time reviewing the content in order to understand why a certain response in a given question is the only response that is correct. This book will help you identify your individual strengths and weakness in each of the 8 reportable subject areas as defined in the Red Seal occupational analysis for this trade.

Who Qualifies to Write the Red Seal Exam?

There are a number of ways to qualify for the Red Seal examination. Before a candidate is allowed to challenge the Heavy Duty Equipment Red Seal Exam, they must have satisfied one of the following requirements:

- Attained satisfactory time and experience working in the heavy equipment trade
- Graduated from a recognized heavy equipment apprenticeship training program, or
- Obtained an out-of-province provincial certificate of qualification from any other province as a heavy duty equipment technician.

A candidate may also be an experienced tradesperson who has not completed an apprenticeship or be an individual with a Canadian military certificate of qualification with a QL5 rating in the trade. The candidate may also have qualifications or experience equivalent to an apprenticeship. For example, individuals in Ontario must complete a Trade Equivalency Assessment (TEA form and Membership Application Form available from the Ontario College of Trades website by email, mail, or in person. In order for the TEA application to be approved, the applicant must provide adequate proof that they have experience and skills in the trade that equal or exceed the Training Standards of an Ontario apprenticeship. Once the forms have been submitted to the college, it takes approximately 8 to 10 weeks for the assessment decision.

In many provinces there are two options for qualifying to write the Red Seal Exam. One is to become a registered heavy equipment apprentice and have followed a prescribed on-the-job training program that consists of completing the on-the-job Training Standard in the Apprenticeship Training Standard Log Book as well as successfully completing all levels of the in-school portion of the prescribed training as identified in the provincial curriculum for this trade. In the latest update to the Ontario Training Standard, Ontario Industries have identified 6,280 hours as the duration necessary for an apprentice to become competent in the skills required to complete their on-the-job apprenticeship training. As a registered apprentice in Ontario, he or she must also have successfully completed 720 hours of in-school training during the course of their apprenticeship at an MTCU (Ministry of Training, Colleges and Universities)-approved school.

The Apprenticeship Training Standard Log Book identifies all the competencies associated with the Heavy Equipment trade in Ontario. It describes how the apprentice must be able to perform each separate skill in order to be considered competent in that skill. The apprentice is responsible for maintaining a training record in the form of an Ontario College of Trades Training Standard Log Book. The sponsors (employers) are required to sign off on each skill as competencies in the trade are achieved. Once a candidate is ready to challenge the exam, they must contact the nearest MTCU office to make a request to write the Red Seal exam. Candidates will be notified once they are approved to challenge the exam and will be provided with a location, time, and date. If an apprentice requires special accommodations due to a disability or langauge barrier, they should notify their regional MTCU office when they apply to write their exam.

Scope of the Heavy Duty Equipment Red Seal Exam

The Canadian Red Seal program has been in existence for over 50 years. It is generally acknowledged that the Red Seal program is the mark of excellence in the Heavy Equipment trade in Canada. The NOA (National Occupational Analysis) is the standard used to develop the interprovincial examinations (Red Seal exams). The exams are used by the provinces to assess the apprentices skills before issuing a Red Seal licence and to ensure the apprentices meet a certain level of excellence. Industry partners across Canada drive the development of the NOA. These trade experts from each province and territory along with curriculum-content writing experts recruited from a college are assembled on a regular basis in Ottawa to develop the Red Seal Standards and exam questions. The Red Seal exam consists entirely of multiple choice questions. Diagnostic questions are the most common type of question on the exam, and many examples of these appear in the two practice exams in this book.

The National Occupational Analysis (NOA and Studying for the Red Seal)

The NOA is used for Red Seal trade exams to define and describe the knowledge, skills, and abilities required by a competent tradesperson. For this reason, the content of the Red Seal exam is based on the NOA. The NOA is, therefore, an excellent tool to use as a study guide for preparing oneself to challenge the Red Seal exam. The NOA document is organized into major subject areas called Blocks. Each of the 8 blocks are divided further into Tasks and Sub-Tasks.

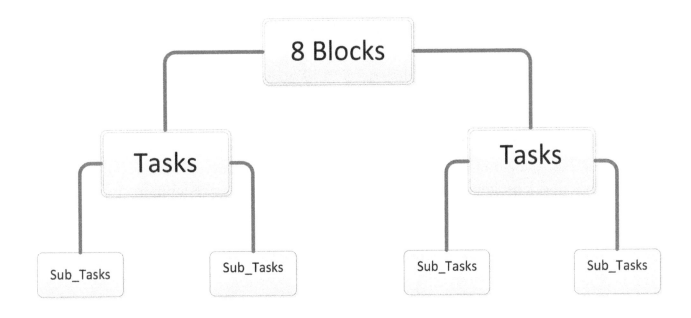

The complete NOA document can be found at **http://www.red-seal.ca/trades/tr.1d.2s_1.3st-eng.html**

It should be printed and used as a checklist of the major and minor skills and you should begin by highlighting areas you feel most competent in. The areas in which your knowledge and understanding need strengthening should be left blank (to be checked once you have studied up on these areas). This procedure ensures that you systematically reinforce all of the weaker areas of your knowledge as you move though the Tasks and Sub-Tasks. Each Sub-Task in the NOA identifies the skill, knowledge, and key competencies that must be performed for each Sub-Task. As such, the Sub-Tasks are the important level of study for the tradesperson to focus on.

How the Training Standard Relates to the NOA
The main difference between the Training Standard and the NOA is that the Training Standard breaks down the competencies that each apprentice must have mastered before he or she is allowed to write the Red Seal Exam. These are created at the provincial level. The NOA, in contrast, breaks down all the trade knowledge into sections and are created at the federal level by all provinces and territories together.

Thus, the Training Standards from each province does reflect approximately 90% of knowledge that will be tested on the C of Q, as do the NOA sections that define the exam content. The major benefit of using the NOA to checkmark your study of weaker areas of your knowledge is that the NOA breakdown of subjects determines how many questions you will get in each Block. This percentage of questions in the NOA will let you know which sections of the trade knowledge will be tested more heavily than others, making it a useful tool for study.

Using the NOA to Study
After acquiring the complete NOA document, print it and review the breakdown of each Block, Task and Sub-Task. Check off any that you are competent in. This list of unchecked Sub-Tasks will serve as a study guide before attempting to challenge the Red Seal Exam. Following these suggestions will ensure the greatest chances of success. Review one Block at a time using the suggested text books to review the line items (these are the Block check list items I use when teaching the Red Seal preparation course) that you are not completely familiar with. This is the key to successfully challenging the Red Seal Exam. I must emphasize the importance of taking the time to prepare to challenge the Red Seal Exam. Too many individuals attempt the Red Seal Exam without sufficient preparation and are disappointed when they aren't successful. The failure rate is quite high in Ontario for first time challengers.

Before you challenge the two sample exams provided in this book, you should identify which areas of the NOA tasks list provided require further understanding (the unchecked Sub-Tasks or Tasks of the NOA). Once the Tasks and Sub-Tasks that require further review are identified, refer to the appropriate text book list supplied later in this chapter to review the content required. The NOA is broken down into 8 Blocks (A to H). Complete one Block at a time before moving on to the next Block. Locate the content chapters required in the three recommended reference text books. Read over the content in question in the appropriate chapters and answer the questions at the end of each chapter before continuing on to the next Block. The first step towards passing your Red Seal exam is to strengthen all the weaker areas of your trade knowledge. Then, the two tests can be taken to measure how effective your study of the material has been and where further study is required.

Understanding the Basics of Trade Knowledge

Many technicians learned their trade skills by learning on the job, and they may not necessarily have gained a deep understanding of the basic theory behind the task. When it came time to write a theoretical multiple choice Red Seal exam, individuals without a strong understanding of the basic theory compared with those with a strong understanding of basic theory are at a clear disadvantage due to their lack of understanding of the basics. Having worked in the Ontario College apprenticeship program for the last 18 years, I have seen hundreds of apprentices graduating from in-school programs without a deep understanding of the basics in each of the 8 reportable subjects in the Ontario apprenticeship curriculum, learning only enough to pass the in-school exams. These exams are based on the in-school curriculum and not on the Training Standards which are the basis for the Red Seal exams, and when it comes time to write the Red Seal Exam, they don't always pass.

I have never forgotten the lesson I learned many years ago as an apprentice. I was working in an overhaul shop about a year into my apprenticeship when my mentor (he was in charge of apprenticeship development) at the time came up to me and asked me a question about the 6-speed transmission I had just rebuilt. I felt pretty confident about the answer because it was around the 10th one I had overhauled in the last three weeks. He pointed to a gear on the countershaft and asked me which gear range was this gear used in in a 6-speed transmission and indicated he would be back in 15 minutes for the answer. I looked at it for quite a while and had to refer to the shop manual before I figured it out. When he returned he asked me if I had to look it up to get the answer — I sheepishly said that I had. He taught me a lesson I never forgot to this day: always be sure you understand how a component works before you start to disassemble any component.

It's one of my favorite questions to pose to apprentices in the labs I teach: how does this component actually work? You would be surprised how many apprentices scramble for this answer. The point here is that the on-the-job Training Standard book that an apprentice uses to track his or her progress reflects the practical hands-on knowledge that the Red Seal tests rather than the in-school theoretical knowledge. There are really four questions you need to ask and have the answer for to be sure your knowledge of a given Sub-Task is adequate in terms of the trade knowledge that the Red Seal tests: 1) what is the component? 2) what does the component do? 3) what does the component consist of? 4) How does the component work each time you perform a task?

The in-school content is based on the apprenticeship provincial in-school curriculum, which makes up approximately 10% of your education as an apprentice. The other 90% is what you learn on the job, which is based on the apprenticeship Training Standard. Therefore, it's easy to see where the majority of a test-taker's study for the exam should be focused. The 10% of knowledge you learn in school provides the foundation for learning and understanding of the practical on-the-job learning the apprentice gains while performing the prescribed Tasks and gaining competencies in the Training Standard. The questions on the Red Seal Exams are not mostly going to be theoretical in nature, but will be application-based or diagnostic in nature.

Some of the NOA sections will require a lot more study and prep time. An example I've often used with apprentices in one of my hydraulic courses is the importance of understanding how a 4-way 3-position control valve with a float centre, open centre or closed centre differ from each other. Without this understanding it becomes difficult to analyze an exam question which describes a particular condition that occurs in a hydraulic circuit that utilizes a float centre 4-way 3-position control valve that does not operate as intended.

Here's a sample question:

Which of the following hydraulic control valve designs would allow the blade of a grader to follow the contours of the ground during grading operation?

    A. Tandem centre.
    B. Vented centre.
    C. Open centre.
    D. Float centre.

The correct response is D.

The float centre envelope is built into the control valve of a grader so that the blade can follow the contours of the roadway during the grading or

> plowing operation. Both sides of the cylinder that control the vertical blade travel is vented to the tank through the control valve allowing the blade to follow the road contours. Answer A (Tandem centre) would not allow the blade to follow the contours of the roadway. This valve lets pump flow go to the tank and blocks both cylinder ports which would prevent the blade from floating. Answer B (Vented centre) does not refer to any known design used in spool type control valve design. Answer C (Open centre) refers to a spool valve that uses all ports in an open centre condition that allows flow to and from all ports including pump flow.

This is a good example of the kind of knowledge you will require on all the Sub-Blocks to succeed on the Red Seal exam test.

After reviewing the Sub-Task checklist in this book, most individuals will realize that they need to get back into the textbooks to review many of the Sub-tasks to gain this deeper understanding of the Sub-Tasks. The questions developed for the Red Seal Exam are almost always diagnostic in nature and will require a deep understanding of the theory to successfully answer the questions. Learning the theoretical rationale behind the question is of great importance; if you have the theoretical rationale, then if the question were to change you could still determine the correct answer. There really is no short cut to this process if you are to be successful.

To complete the suggested course of study recommended in the Sub-Task check list, 6 textbooks are recommended for this purpose and should be used to review material that the tradesperson may not be totally familiar with.

Recommended Textbooks

ISBN-10-1-305-07362-2	Heavy Duty Truck Systems 6e by Sean Bennett
ISBN-10-1-133-69336-9	Heavy Equipment Systems 2e by Robert Huzij, Angelo Spano, Sean Bennett
ISBN-10-1-111-64569-8	Medium/Heavy Duty Truck Engines 4e by Sean Bennett
ISBN-10-1-4018-801304	Electricity and Electronics by Sean Bennett
ISBN-10-1-4180-1372-2	Brakes, Steering and Suspension Systems by Sean Bennett
ISBN-10-1-4018-0950-4	Heating, Ventilation and Air Conditioning by John Dixon

How to Use the Practice Exams

Multiple choice exams can be difficult to answer correctly. Often there may be two answers that may appear to be correct. The importance of understanding the question becomes extremly critical to success. When you challenge the first exam be sure not to guess at the answer; you have only a 25% chance of guessing correctly. Only answer questions you are confident you know the answers to. Once this first exam has been completed, focus on the questions and subject matter you did not understand. Review the rationale for each question you did not answer correctly. If that is not enough for you to clearly understand why a certain response is correct and all the others are not, you need to look up the subject matter in the recommended textbooks and use Google or YouTube to review the subject matter before retaking the exam again. Once all the necessary content has been reviewed and you are confident that the subject matter for the questions you did not answer successfully has been reviewed, redo the exam for the second time. If you didnt achieve at least 90%, then you need to repeat the same process as outlined above again untill you can achieve a consistent 90%.

    The Red Seal exam bank has hundreds of questions which may be selected for the one exam you will need to pass. The pass rate for successfully completing the Red Seal exam is not high. I cannot over-emphasize the importance of thoroughly understanding the stem (the actual question) content. The exam questions have been written in such a fashion that they are mostly diagnostic in nature, just like the Red Seal exam questions. These questions have been written to address the practical application of diagnostics and repair procedures that a typical technician would encouter on the job.

    Not everyone learns the same way; its important to know what works for you. I recommend that you complete the first exam and take a day or two off before coming back to review the questions that you have not answered successfully. Craming does not work; there is just to much material to absorb. In my experience teaching over the years, I have been led to the conclusion that memorizing the answers does not work effectively. You must truly understand the subject matter thouroughly. I often have discussions with my students in the classsroom about the difference between knowing something and thorougly understanding the subject. The example I often use involves how a 5-pin relay works. Most of my students will know what function pin 85 and 86 serves on a 5-pin relay. But when I ask them what would occur if there was a short in the coil between pin 85 and 86, I often get a confused response. Just knowing what each pin on a relay is for is not enough to answer a diagnostic type question on this type of relay. Understanding how it works and what occurs when it does not work correctly is of the utmost importance especially when it comes to the Red Seal Exam. If you dont know the answer to this question then you too need to explore this subject matter in greater depth.

Using the Practice Exams and Exam Profiles ➤ 2

To correctly answer a multiple choice question, one must consider the stem (the question) and chose from among four distractors (the possible answers to the question) the correct answer. There will never be two correct answers. First read the question carefully to be sure you understand what is being asked. Many individuals don't analyze the question correctly and misinterpret what is actually being asked. The following question will illustrates the proper way to analyse a multiple choice question.

Example of a Red Seal Question

A technician has diagnosed a brake performance problem on a loader equipped with hydro-boost brakes. During a visual inspection, he has found a leak that appears to be coming from the rear of the master cylinder. Which one of the following is the source of the leak?

    A. Internal leak in the master cylinder.
    B. Internal leak in the brake booster.
    C. Leak at the secondary seal.
    D. Leak at the primary seal.

The first step is to read the question carefully and determine what is actually being asked. First, establish that you are looking for "what is occuring" and "where is it occuring." The "what" is a leak. The "where" is the rear of the master cylinder. If the answer is not obvious to you, then the correct study approach to use is to refer to and review the hydraulic brake chapter in the appropriate text book that describes the operation of this master cylinder. This takes us back to the importance of the in-school portion of your apprenticeship where you would have had the opportunity to learn the theory of how this master cylinder actually works, and what could occur if it didn't work correctly.

The importance of understanding this theory of operation of the master cylinder becomes quite evident here. The stem of this question could be changed to make anyone of the responses correct just by changing a few words in the question.

Example: a technician has diagnosed a brake performance problem on a loader equipped with hydro-boost brakes. During a visual inspection, he has found a leak that appears to be coming from the *front* of the master cylinder. Which one of the following is the source of the leak?

   A. Internal leak in the master cylinder.
   B. Internal leak in the brake booster.
   C. Leak at the secondary seal.
   D. Leak at the primary seal.

Answer C is correct.

Explanation:

When the brakes are activated, the primary piston is forced to move forward thereby closing the inlet port, which allows brake pressure to build in the primary circuit. The increase in primary brake pressure forces the secondary piston to move forward, closing off the inlet port and causing the pressure in the secondary circuit to increase. When the brakes are released the pressure in the brake circuit drops instantly as the pistons retract. The compensating ports located in the piston chambers help vent the brake fluid back into the reservoirs. Note that the single bore houses both the primary and secondary pistons for both circuits. They are separated by seals, which prevent one circuit's fluid from entering the other circuit. A clear understanding of how the master cylinder works allows you to analyze the question and responses and rule out A and B as these would not result in an external leak. Response D would not result in an external leak but would leak internally. So the only response that could be correct would be C. If the wording in the stem was changed to "Front" from "Rear" to that same question it would change the correct answer to A. So you can see the importance of reading the question carefully and understanding what is being asked. A simple one-word change to the question can change the question entirely.

Taxonomy Levels for Exam Questions

Taxonomy levels are a system for categorizing questions based on their difficulty, or the extent and depth of knowledge required to answer them correctly. A test taker will face many questions within each of the taxonomy levels on the exam, many requiring in-depth understanding of the material from the Sub-Task items. The questions may be developed using three taxonomy levels: level one, two, and three. Level one questions test your ability to recall and understand definitions, formulas and principles. Level two questions test your ability to apply knowledge of the procedures to a given situation. Level three questions test your ablity to interpret data, solve problems, and arrive at valid conclusions.

The questions below show examples of the three different taxonomy levels that may be used in a typical exam question.

Example of Taxonomy Level 1 Question
Which of the following is used to protect bulldozers electrical circuits against electrical overloads?

- A. A zener diode.
- B. A diode bridge.
- C. A circuit breaker.
- D. A NPN transistor.

The first example above is straight recall question—you either know the answer or you don't. This is taxonomy level 1.

Example of Taxonomy Level 2 Question
List the three basic principles in the correct order of ECM operation.

- A. Input, Processing, Actuation.
- B. Input, Processing, Output.
- C. Input, Processing, Feedback.
- D. Input, Processing, Regulation.

The second example tests your ability to use knowledge of the procedures as applied to a given situation—a taxonomy level 2 question.

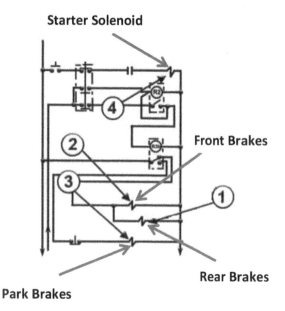

On a loader with the engine running and the hydraulically applied brakes released, which solenoid is energized on the schematic?

A. Solenoid No. 1.
B. Solenoid No. 2.
C. Solenoid No. 3.
D. Solenoid No. 4.

The third type of question is a taxonomy level 3 question which requires you to be able to read the schematic and assess which solenoid is energized. This is a critical-thinking type of question which requires one to understand how the brake system works.

The different difficulty levels of the questions should underscore the importance of thoroughly studying the list of Sub-Tasks, ensuring your knowledge is deep enough for you to be able to answer a taxonomy 3 question drawn from any of the Sub-Tasks.

Study Planning and Tips

In order to be successful at your first attempt at aquiring your Red Seal certification, you must organize your study procedures. The best method is to develop a study plan. This involves assessing the areas of weakness you need to strengthen and setting aside enough time to allow you to study the content you identify from the Sub-Tasks checklist. Prepare a weekly study plan outlining your goals for the week based on the Sub-Task checklist. You should identify two catagories for study purposes: Sub-Tasks you only need to review, and Sub-Tasks that require more in-depth study. Create a timeline for yourself working back from your exam date so that you have adequate time to study each Sub-Task. Each person is different and will have different levels of knowledge in different areas of their trade knowledge at the time that they begin their study. Some individuals I have helped through this process only required a month to prepare; others required over three months.

Some of the Sub-Tasks can be further researched on YouTube or online as well as in the recommended text books. In some cases the Sub-Tasks in question may not be readily avaliable for you to review on the job. For example, if you worked for a employer that did not perform any air conditioning repairs onsite you would not get any experience in this area. There may be a number of areas on which you are not able to gain pratical experience, so you must rely on extensively studing these areas in greater length using other resources. Many equipment manufacturers and schools have YouTube online resources available for review at no cost. Many companies these days contract out specialized work to companies that speciliaze in specific tasks as in the example I mentioned air conditioning repair. The thing to keep in mind is that this Red Seal exam is a theory exam, so one could learn how Air Conditioning works and still be able to answer questions correctly without actually having worked on the system. Make a list of Sub-Tasks that you are not able to review on the job and look them up online on YouTube if you can.

As I have said, candidates who fail the Red Seal exam generally don't understand the material well enough to analyse the questions properly to enable them to see what the question is really asking, and that is the most common reason for failure. The first-time failure rates are quite high, generally caused by lack of proper preperation. Many candidates simply feel they have enough knowledge to pass the first time. There is no substitute for well-organized preparation. Remember, the higher the taxonomy level of the question, the more important it is to analyze the parts of the question correctly. The exam questions are often misintepreted, so read the questions and all the possible answers carefully, and eliminate the obvious distractors.

Don't spend an excessive amount of time on one question; instead come back to it later. Always answer every question even if you have to guess. You have a 25% chance of guessing the correct answer, and if you can eliminate a distractor or two your chances go up even higher. Be careful when considering changing your original answers. Past history has shown that the correct answers are often replaced by incorrect answers.

Using the Exam Profile to Weight Your Study Time

The NOA chart shows the breakdown in the form of a pie chart which breaks down the imate number of questions from each block. For example, 9% of the questions on the Heavy Duty Equipment Technician Red Seal Exam will be based on Block A.

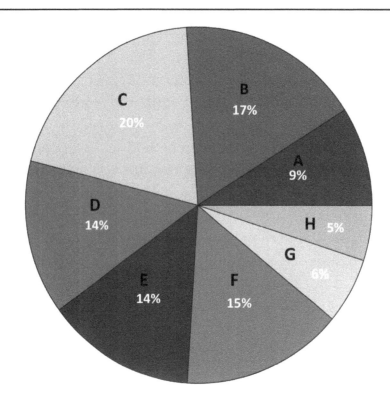

Block A Occupational Skills 9%

Block B Engine and Engine Support Systems 17%

Block C Hydraulic, Hydrostatic, and Pneumatic Systems 20%

Block D Drive Train Systems 14%

Block E Steering, Suspension, Brake Systems, Wheel Assemblies, Undercarriage 14%

Block F Electrical and Vehicle Managment Systems 15%

Block G Enviromental Control Systems 6%

Block H Structural Components, Accessories, and Attachments 5%

Red Seal Exam Breakdown:

The Heavy Duty Equipment Technician Red Seal Exam currently has a total of 135 questions. The exact number of questions is shown in the exam breakdown that follows. The exact number of questions has been held to 135 at present but can change in the future.

Our practice exams number only 100 questions, but the weightings of each reportable subject are the same as that of the Red Seal exam.

		Questions
Block A	Common Occupational Skill 9%	12
Task 1	Uses and Maintains Tools and Equipment	
Task 2	Performs General Maintenance and Inspections	
Task 3	Organizes Work	
Task 4	Performs Routine Trade Activities	
Block B	Engines and Engine Support Systems 17%	23
Task 5	Diagnoses Engine and Engine Support Systems	
Task 6	Repairs Engines and Engine Support Systems	
Block C	Hydraulic, Hydrostatic and Pneumatic Systems 20%	27
Task 7	Diagnoses Hydraulic, Hydrostatic and Pneumatic Systems	
Task 8	Repairs Hydraulic, Hydrostatic and Pneumatic Systems	
Block D	Drivetrain Systems 14%	19
Task 9	Diagnoses Drivetrain Systems	
Task 10	Repairs Drivetrain Systems	
Block E	Diagnoses Steering, Suspension, Brake Systems, Wheel Assemblies and Undercarriage 14%	19
Task 11	Diagnoses Steering, Suspension, Brake Systems, Wheel Assemblies and Undercarriage	
Task 12	Repairs Steering, Suspension, Brake Systems, Wheel Assemblies and Undercarriage	
Block F	Electrical and Vehicle Managment Systems 15%	20
Task 13	Diagnoses Electrical Systems	
Task 14	Repairs Electrical Systems	
Task 15	Diagnoses Electronic Vehicle Managment Systems	
Task 16	Repairs Electronic Vehicle Managment Systems	
Block G	Enviromental Control Systems 6%	8
Task 17	Diagnoses Enviromental Control Systems	
Task 18	Repairs Enviromental Control Systems	
Block H	Structural Components, Accessories and Attachments 5%	7
Task 19	Diagnoses Structural Components, Accessories and Attachments	
Task 20	Repairs Structural Components, Accessories and Attachments	
	Total Number Of Question	135

The table used here is based on the NOA weighting applied to each of the 8 Blocks with the number of questions that are asked from each area of the NOA analysis reflecting what a typical tradeperson should know.

Acknowledgements:

I would like to thank the following individuals who helped with the final editing to the questions used in this study guide.

- Jim Carvel (Professor, Heavy Equipment Program Cambrian College)
- Ken Knoll (Professor, Heavy Equipment Program Cambrian College)
- Arthur Pringle (Professor, Heavy Equipment Program Cambrian College)
- Mike Lahaie (Professor, Heavy Equipment Program Cambrian College)

Practice Exam 1 ➤ 3

The following practice exam has 100 questions and follows the weighting of the Red Seal exam. The answer explanations can be found on page 65.

Block A. Common Occupation Skills (9 Questions)

1. A technician purchased an aerosol cleaner for use in the shop he works in. The label on the product says "MSDS available" on the manufacture's website. What does the Acronym "MSDS" refer to as defined in the Workplace Hazardous Material Information System (WHMIS)?

    A. Material Safety Data Service.
    B. Main Safety Data Sheet.
    C. Material Safety Data Sheet.
    D. Management Shop Data System.

2. During a pre-work inspection, a technician finds cracks in the insulation of one of the welding cables he is about to use. What repair must be performed before he begins welding?

    A. Repair the cracks using only electrical tape.
    B. Mark for repair after welding is completed.
    C. None. Breaks will not affect the weld for now.
    D. Before welding, replace the cable with cracks.

3. A technician is checking the run-out of a flywheel housing with a dial indicator. During a single rotation of the flywheel, the reading on the positive side of the zero peaks at 0.002" (0.05 mm) while the reading on the negative side peaks at 0.005" (0.13 mm). What is the total indicated run-out (TIR)?

   A. 0.003" (.08 mm)
   B. 0.005" (.13 mm)
   (C.) 0.007" (.18 mm)
   D. 0.010" (.25 mm)

4. A technician is about to calibrate a standard 0"–1" micrometer before he measures a shim. How many thousands of an inch are there in one revolution of the thimble?

   A. 0.010"
   (B.) 0.025"
   C. 0.050"
   D. 0.100"

5. A technician is performing a leak test on an oxy-acetylene system before doing any burning or welding. Which of the following is a standard procedure and must be followed?

   A. Apply antifreeze and look for evidence of bubbles.
   (B.) Apply a mixture of soap and water and look for bubbles.
   C. Apply an inert gas and look for discoloration.
   D. Apply clean machine oil to locate the leak.

6. A technician is about to perform gas welding on a 3–12 mm (1/8"–1/2") Material. What lens shade number should be used?

   A. 2 or 3.
   (B.) 5 or 6.
   C. 8 or 9.
   D. 11 or 12.

7. A technician is using a wire rope sling to lift a heavy load. What is the safety factor that is generally used for this type of lifting device?

   A. 2 to 1.
   B. 3 to 1.
   C. 4 to 1.
   (D.) 5 to 1.

8. A technician is installing a crankshaft lip seal into the front housing of an engine. Which of the following is part of the correct installation procedure?

    A. The open side of the lip should face inwards.
    B. The closed side of the lip should face the lubricant.
    C. Grease should be applied to the seal housing.
    D. Silicone grease should be applied to the seal lip.

9. A technician is about to use the torque-to-yield technique on fasteners. Which of the following procedures best describes this?

    A. Re-torque all fasteners to specifications after 100 hours of use.
    B. Clean and install fasteners and dry torque fasteners to specifications.
    C. Rotate a specified number of degrees and then torque to specification.
    D. Torque to preload specification and then rotate a specified number of degrees.

Block B. Engines and Engine Support Systems (17 Questions)

10. When the operator performs his pre-operational checks in the morning, he finds that the top radiator hose has collapsed. Which of the following can lead to this condition?

    A. The thermostat needs to be replaced.
    B. The coolant viscosity is too high.
    C. The radiator cap is defective.
    D. The top radiator hose needs to be replaced.

11. A technician has found that the radiator cap is failing to seal properly on a diesel engine and does not hold pressure in the radiator. Which of the following would be more likely to occur?

    A. Lowers coolant's boiling point.
    B. Lowers operating temperatures.
    C. Increases hydrocarbon emissions.
    D. Increases operating temperatures.

12. A technician is inspecting the crankshaft main journal cheek to the rod journal area and finds radiating cracks have developed in this area. Which of the following could be responsible for these cracks?

    A. Torsional stress in the crankshaft.
    B. Misalignment of the main bearing bores.
    C. Crankshaft lubrication failure.
    D. Bent connecting rod.

13. A technician tests the inlet restriction of a turbo-charged diesel engine at full engine load. What is generally considered the maximum restriction allowed on the suction side of the intake?

    A. 25" of mercury.
    B. 25" of water.
    C. 25 psi.
    D. 25 KPa.

14. A technician checks a suspected bad injector for back leakage and finds the back leakage is higher than normal. Which of the following is likely responsible for this condition?

    A. Internal nozzle wear.
    B. Low cracking pressure.
    C. High cracking pressure.
    D. Sticking nozzle assembly.

15. A technician replaces a high-pressure injector line on a diesel engine equipped with an inline injection pump with one that is shorter in length. What affects will this likely have on the operation of the engine?

    A. The timing will be advanced slightly in that cylinder.
    B. The timing will be retarded slightly in that cylinder.
    C. That injector will have slightly lower cracking pressure.
    D. That injector will have slightly higher cracking pressure.

16. A technician is testing a loader for excessive exhaust particulate. Which of the following emission test instruments uses a light extinction principle to measure particulate?

    A. VOC (volatile organic compound analyzer).
    B. Seven-gas analyzer.
    C. Two-gas analyzer.
    D. Opacity meter.

17. When an emission test is conducted on a diesel engine that is operating at below normal coolant temperatures, which emissions would likely be elevated due to the low engine operating temperatures?

    A. Oxides of nitrogen.
    B. Carbon dioxide.
    C. Nitrous oxides.
    (D.) Hydrocarbons.

18. A technician is checking the operation of catalytic converter on a diesel engine. Which of the following emissions does a catalytic converter help control?

    A. Nitrogen and nitrogen oxides (NO_x) is burned.
    (B.) Carbon Monoxide (CO) is converted to carbon dioxide (CO_2).
    C. Hydrocarbons (are filtered out).
    D. Particulate matter (PM) is filtered out.

19. A technician is going to weld on a loader equipped with electronic engine controls. Which of the following should be performed before any welding takes place?

    (A) Disconnect both negative and positive battery cables.
    B. Disconnect the battery negative cable only.
    C. Disconnect the battery positive cable only.
    D. Disconnect the computer connections only.

20. A technician is testing the operation of an aneroid valve on a turbo-charged diesel engine. Which of the following does the aneroid valve control?

    (A) Limits fuel delivery during acceleration to minimize black smoke.
    B. Increases fuel pressure until the engine speed starts to increase.
    C. Is only used at high elevations to compensate for lower air pressure.
    D. Increases turbo-boost pressure during acceleration to minimize smoke.

21. What will occur when water enters the fuel circuit of a diesel engine with a secondary fuel filter that is capable of entrapping particulate as small as two microns?

    (A) The fuel filter will become plugged and reduce fuel flow.
    B. The fuel filter will likely collapse from injector suction.
    C. The water will be absorbed into the fuel and burned off.
    D. The filter media will break down and eventually collapse.

22. A technician is performing a cylinder leakdown test on a diesel engine and notices that air is coming out from the dipstick tube during the test. Which of the following could be responsible for this?

   A. Excessive wear on valve guides.
   B. Excessive wear on the intake valve seats.
   C. Excessive wear on the piston rings and liners.
   D. Excessive wear on exhaust valve seats.

23. A technician is performing a bearing leakdown test on a diesel engine. What would excessive oil leakage from one connecting rod-bearing journal area indicate?

   A. Excessively worn connecting rod bearing.
   B. Excessive main bearing endplay.
   C. Excessive rod-bearing crush.
   D. Insufficient rod journal side clearance.

24. After a technician has replaced a turbocharger on a diesel engine, which of the following should be performed before the initial startup?

   A. A cylinder leakage test must be performed.
   B. The turbocharger must be primed with oil.
   C. Lube oil pressure test must be performed.
   D. A cylinder compression test must be performed.

25. A technician installs a new 15 psi (1.1 Bar) pressure cap on a diesel engine radiator. What effect would this have on the coolant's boiling point?

   A. Raise it by 25°C (45°F).
   B. Lower it by 25°C (45°F).
   C. Raise it by 8.3°C (15°F).
   D. Lower it by 8.3°C (15°F).

26. A technician is going to perform an efficiency test on a turbocharged diesel engine intercooler. Which of the following procedures would he use?

   A. Measure the air temperature drop across the inlet and outlet under load.
   B. Measure the air pressure drop across the inlet and outlet at low idle.
   C. Compare boost pressure readings between full load and high idle.
   D. Measure the inlet restriction from one side to the other with a manometer.

Block C. Hydraulic, Hydrostatic and Pneumatic Systems (20 Questions)

27. A fixed displacement gear pump is diagnosed with low oil flow in a hydraulic system. Which of the following would likely be responsible for reduced oil flow from the gear pump?

    A. Excessive backpressure in the tank.
    B. Excessive internal pump wear.
    C. Pump rpm has been too high.
    D. Relief valve setting set too high.

28. A technician is checking the operation of a priority flow control valve on a loader. Which of the following should he check first?

    A. Ensure minimum pressure in the secondary circuit first.
    B. Check to insure both circuits are fed at the same time.
    C. Ensure maximum pressure in the secondary circuit first.
    D. Ensure the primary circuit is supplied before the secondary circuit.

29. A technician checks the bladder-type accumulator pressure on a loader's brake circuit and finds the gas charge pressure is too high in the accumulator. Which of the following could result from this condition?

    A. Excessive brake pressure will occur in the wheel end.
    B. Lower hydraulic pressure will occur in the system.
    C. Damage to the poppet assembly and/or the bladder.
    D. Greater oil volume will be stored in the system.

30. After a technician replaces a single acting telescoping hydraulic cylinder on a drill rig, which of the following procedures must be performed before the equipment is returned to service?

    A. Air must be bled from the port relief valve.
    B. Air must be bled at the main control valve.
    C. Air must be bled from the telescoping cylinder.
    D. Air must be bled from the hydraulic pump first.

31. A technician is diagnosing a suspected hydraulic system low-pressure problem. Which of the following criteria must be met before proceeding?

    A. Start and operate the equipment at low idle for 10 minutes.
    B. Ensure that the hydraulic oil cooler is bypassed.
    C. Test that the hydraulic pump is not leaking internally.
    D. Check that the hydraulic system is at normal operating temperature.

32. A technician finds that the boom circuit of a loader drifts down on occasion. Which of the following would likely be responsible for this?

    A. Leak at the rod end of the dump cylinder.
    B. Intermittent sticking of the main relief valve.
    C. The dump/hoist control valve is sticking.
    D. Excessive wear to the hydraulic pump gears.

33. An operator complains that the blade on his dozer is constantly drifting. Which of the following would likely be responsible for this condition?

    A. Blade cylinder mounting pins worn excessively.
    B. Blade counterbalance valve is sticking.
    C. Bypassing in the blade cylinder packing.
    D. Plugged strainer causing aeration of the oil.

34. A technician disassembles a gear pump removed from a loader and finds pitting on the thrust plates. Which of the following would likely be responsible for this condition?

    A. Oil aeration.
    B. Oil emulsification.
    C. Excessive gear end play in the pump.
    D. Restriction in the pump pressure line.

35. A technician has diagnosed that the charge pump in a closed centre hydraulic system is cycling too frequently. Which of the following could be responsible for this symptom?

    A. The relief valve is set too low.
    B. A restriction in pump inlet.
    C. Internal charge pump wear.
    D. An internal leak in the circuit.

36. An operator complains that the boom on his loader has a jerky motion to it when he is lifting the boom to dump his bucket. Which of the following is likely responsible for this symptom?

    A. A low oil level in the reservoir.
    B. The line relief is defective.
    C. Cylinder packing is bypassing.
    D. The boom spool valve is defective.

37. A technician has replaced the main relief valve in a hydraulic manifold block on a loader. What must be done before the equipment is started for the first time?

    A. Back off the relief valve adjustment before starting.
    B. Bleed the air from the pressure lines before start-up.
    C. Replace the suction filter before start-up.
    D. Be sure the tank has a vent line connected.

38. A technician needs to reduce the output speed of a variable displacement drive motor. Which of the following could be used to accomplish this task?

    A. Increase the motor displacement by adjusting the swash plate.
    B. Reduce the swash plate angle to zero degrees.
    C. Reduce motor displacement by reducing the swash plate angle.
    D. Increase the variable displacement pump output.

39. After replacing a leaking hose on a loader, the technician notices oil leaking from the top of the hydraulic tank during the initial start-up inspection. Which of the following could cause this to occur?

    A. Main relief valve is defective.
    B. Hydraulic inlet filter is plugged.
    C. The hydraulic reservoir is over-filled.
    D. Hydraulic pump speed is excessive.

40. A technician diagnosing a hydraulic piston accumulator on a loader that has too low a nitrogen gas charge in the accumulator. Which of the following symptoms would likely occur?

    A. Low hydraulic charge pressure.
    B. High oil temperature during cycling operation.
    C. Excessive charge pump cycling.
    D. Excessive volume of oil circulating in the system.

41. A technician discovers a pin hole in the inlet line above the fluid level in a reservoir of a fixed displacement pump. What symptom would occur during operation?

   A. Oil foaming.
   B. Inlet hose collapse.
   C. Oil starvation.
   D. Emulsification.

42. A technician diagnoses a closed-centre hydraulic system on a skid steer loader that is cycling too frequently. Which of the following would likely be responsible for this?

   A. A defective relief valve.
   B. A restriction at the pump inlet.
   C. An internal leak in the circuit.
   D. A defective flow divider.

43. A technician diagnoses a cavitation condition that is occurring in a hydraulic system with a fixed displacement pump. Which of the following could be responsible for this symptom?

   A. Pump operating at too slow of an rpm.
   B. Excessive positive head pressure.
   C. Low oil level in the hydraulic tank.
   D. Excessively high oil temperatures.

44. A technician is performing a vacuum restriction test at the pump inlet of a hydraulic pump. Which of the following readings would be considered the maximum allowed?

   A. 127 mm (5″) of H_2O.
   B. 127 mm (5″) of Hg.
   C. 736 mm (29″) of H_2O.
   D. 736 mm (29″) of Hg.

45. A technician has diagnosed that the piston seals are leaking internally on the boom cylinders of a loader. Which of the following symptoms will occur during operation of the loader?

   A. Raise cycle time will become longer.
   B. Lowering cycle time is shorter.
   C. Load lifting capability is increased.
   D. Boom cylinder operation is erratic.

46. A technician records a hydraulic pressure reading of 800 psi when the boom of a loader is extended to full stroke. When the system pressure is checked with the boom resting on its stops, the pressure is a normal 1,800 psi. Which of the following could cause this problem to occur?

   A. Port relief defective.
   B. A bent cylinder rod.
   C. Hydraulic strainer restriction.
   D. A bent spool valve.

Block D. Drivetrain Systems (14 Questions)

47. A technician is performing a converter stall test on a loader with a countershaft power shift transmission that shows an rpm reading that is higher than normal stall rpm. Which of the following could cause this condition?

   A. Forward clutch slippage.
   B. Converter-in pressure low.
   C. Converter lockup slipping.
   D. Converter-out pressure high.

48. A technician is performing a converter stall test on loader with a power shift countershaft transmission. Which of the following is the correct procedure?

   A. Brakes applied and engine rpm under full load during turbine stall.
   B. Brake applied and engine rpm with the highest possible turbine speed.
   C. Brakes released and high idle with no load and zero turbine speed.
   D. Brakes released and high idle with half load and turbine at maximum speed.

49. The operator of a haulage truck has indicated that they hear an unusual noise coming from the rear of the truck whenever they are cornering. Which of the following would be most likely to be responsible for this noise?

   A. Excessively worn differential side gears.
   B. Too heavy toe contact on the ring gear.
   C. A couple of chipped ring gear teeth.
   D. Excessive backlash between the pinion and crown.

50. A technician has dismantled a differential from a line haulage truck and has noticed that the bevel pinion gears have an unusually high amount of wear. Which of the following conditions could be responsible for this?

    A. Excessive reversing under load.
    B. Unequal tire size on the same axle.
    C. Lugging in straight line driving.
    D. Seized differential lock.

51. A technician is overhauling a differential from a loader and is about to adjust the thrust bolt. Which of the following would this adjustment offset?

    A. Minimizes the effects of pinion gear run-out.
    B. Prevents excessive ring gear deflection.
    C. Minimizes the effect of thrust loading the pinion.
    D. Prevents shock loading of the ring gear.

52. A technician is performing a converter stall test on a loader with a power shift transmission. He observes that there is high-speed driveshaft slippage in 2nd gear forward but not in 2nd gear reverse. Which of the following would likely cause this symptom?

    A. 2nd gear clutch-pack discs worn excessively.
    B. Reverse clutch-pack discs worn excessively.
    C. Forward clutch-pack discs worn excessively.
    D. Clutch regulating valve spring is weak.

53. A technician is about to perform a pressure differential test on the loader's power train circuit to locate the source of power train overheating. Which component should he perform a pressure differential test on first?

    A. Torque converter circuit.
    B. Oil cooler circuit.
    C. Oil lubrication circuit.
    D. Clutch pressure circuit.

54. A technician is going to diagnose a bucket loading performance problem on a loader's power shift transmission. Which of the following pressure differential tests should he perform?

    A. Lube cooler circuit.
    B. Clutch pressure circuit.
    C. Case drain pressure circuit.
    D. Torque converter pressure.

55. A technician is diagnosing the source of overheating on a power shift transmission circuit in a 4-yard loader. Which of the following should be checked for first?

    A. Clutch pack or brakes dragging.
    B. Transmission cooler temperature differential.
    C. High clutch pack pressures.
    D. Transmission oil filters restriction.

56. A technician is about to order a drive chain for a grader. Which of the following check would verify that he will order the correct chain pitch?

    A. The distance measured from pin centre to pin centre of one link.
    B. The distance measured from pin centre to pin centre of two links.
    C. The distance measured from pin centre to pin centre of three links.
    D. The distance measured from pin centre to pin centre of four links.

57. A technician has diagnosed excessive differential gear wear during an overhaul. He notices that the teeth on the ring gear show signs of spalling, but the teeth on the pinion have almost no signs of spalling. Which of the following must be performed?

    A. Replace the ring gear and pinion as a set.
    B. Replace the ring gear as well as the side bearings.
    C. Replacing the ring gear is all that is required.
    D. Reuse the gear set as spalling is considered normal wear.

58. A technician is overhauling a front differential from a 4-yard loader and must set the backlash next. What is this setting referring to?

    A. The difference in shimming between the back bearing and the front bearing.
    B. The clearance measured between the pinion teeth and the crown gear teeth.
    C. The preload on the carrier bearings after they are preloaded.
    D. The shimming between the bearings on the pinion cage.

59. A technician is diagnosing a complaint that the power shift transmission on a loader is slow to shift into gear when the operator selects a gear range. Which of the following could be responsible for this symptom?

    A. Worn transmission oil pump is causing low pressure.
    B. Clutch pressure regulator valve is sticking.
    C. Lube pressure regulator valve in the transmission is stuck.
    D. Low converter-out pressure at the transmission cover.

60. A technician is diagnosing a suspected worn charge pump on a loader's power shift transmission circuit. Which of the following symptoms would indicate an excessively worn transmission supply pump?

    A. High lubrication pressure at low idle.
    B. Low reverse clutch pressure at low idle.
    C. Low lubrication pressure at low idle.
    D. Low forward clutch pressure at low idle.

Block E. Steering, Suspension, Brake Systems, Wheel End and Undercarriage (14 Questions)

61. A technician is adjusting the steering knuckles vertical clearance on the front end of a haulage truck. Which of the following adjustments would he be performing?

    A. Adding or subtracting shims.
    B. Selecting the correct draw key.
    C. Adjusting a stop bolt and jam nut.
    D. Adjusting the toe-out dimension.

62. A technician is performing a wheel alignment on a truck and notices that the front tires have a featheredged wear pattern on the treads. Which of the following would likely have caused this wear to occur?

    A. Incorrect toe setting.
    B. Incorrect camber setting.
    C. Incorrect caster setting.
    D. High tire pressure.

63. A technician is performing a front-end inspection on a haulage truck. He notices that the front tires are showing excessive wear on both outside edges. Which of the following would likely be responsible for this condition?

    A. Worn suspension parts can cause constant toe dimension changes.
    B. Excessive negative camber on both wheels cause toe dimension changes.
    C. Excessive toe-out adjustment on both wheels cause toe dimension changes.
    D. Too much positive caster settings on one side only cause toe dimension changes.

64. A technician is changing the rear suspension strut on an off-road haulage truck; the strut fastens to the frame on one side. What would he fasten the other side to?

    A. Torque rod.
    B. King pin.
    C. Cylinder rod.
    D. Equalizer beam.

65. A technician is installing a new accumulator on the ride control system of a 200-tonne truck. Which of the following must be performed before the new accumulator is installed?

    A. Fill the accumulator with oil.
    B. Release the pressure in the accumulator.
    C. Purge the accumulator of all air.
    D. Confirm hydraulic pressure is present.

66. A technician is installing a straight kingpin on a haulage truck's steering knuckle. Which of the following components must be installed to secure it to the steering knuckle?

    A. Snap ring.
    B. Cotter pin.
    C. Draw key.
    D. Lock ring.

67. A technician is checking the rear suspension cylinders on a 200-tonne haulage truck. Which of the following would indicate servicing is required?

    A. The ride height between cylinders varies more than 0.250" (6.35 mm).
    B. The cylinder pressure varies more than 50 psi (345 KPa between cylinders).
    C. The cylinder pressure varies more than 100 psi (690 KPa between cylinders).
    D. The struts don't extend fully when the truck box is empty.

68. A technician is performing a front-wheel alignment on a haulage truck's front end and finds that the left front wheel has excessive negative camber. Which one of following could be responsible for this condition?

    A. Excessive wear in the truck's rear spring shackles and pins.
    B. The knuckle pin bushings on the left side of the axle have excessive play.
    C. The bearing is seized between the steering knuckle and the axle eye.
    D. Incorrect amount of shims between the left spring and the axle.

69. A technician is replacing the friction discs on an internal spring-applied wheel end assembly on a loader. Which precaution must be observed before disassembly of the wheel end?

   A. Brake pressure must be bled off at the wheel ends before removing the housing.
   B. Spring force must be caged mechanically during the removal of friction discs.
   C. Hydraulic pressure must be used to compress the brake application springs.
   D. Accumulator pressure must be bled off to release the brake springs.

70. A technician must adjust the brake charge valve pressure on a loader with inboard internal hydraulically applied brakes. Which of the following brake pressures are they actually going to adjust?

   A. System cut-out.
   B. Brake cut-in.
   C. Service brake.
   D. Accumulator charge.

71. A technician is checking what occurs when the brake cut-in pressure is reached on a loader equipped with hydraulic wet disc brakes. Which of the following describes what occurs?

   A. Wheel-end brake pressure is amplified during operation.
   B. The brake pump will begin to charge the accumulator circuit
   C. Accumulators have reached maximum operating pressure.
   D. The accumulator remains at minimum operating pressure.

72. A technician is diagnosing a slow brake application symptom on a loader with internal spring-applied hydraulically released brakes. Which of the following could be responsible for this condition?

   A. Restriction in the foot brake valve return line.
   B. Restriction in the foot brake valve pressure line.
   C. The accumulator has a low gas charge pressure.
   D. An internal leak in the charge valve assembly.

73. A technician is replacing a shuttle valve in the brake circuit of a loader with internal hydraulically applied brakes. Which of the following would be the correct description of the shuttle valve's function?

   A. Allow the accumulator to charge two circuits at the same time.
   B. Provide a path for oil to reach two components in a series.
   C. Provide two separate paths for oil to flow to the wheel ends.
   D. Direct oil flow to the sequence valve and the wheel end.

74. A technician has diagnosed a low brake pressure problem on a loader with hydraulically applied internal wet disc brakes. Which one of the following could be responsible for this problem?

   A. Excessive brake pedal travel.
   B. Restricted brake pedal travel.
   C. Worn disc brake lining.
   D. Broken disc return springs.

Block F. Electrical and Vehicle Management Systems (15 Questions)

75. A technician is diagnosing a problem in an electrical lighting circuit on a loader; he discovers that the rear lights are getting back-fed. Which of the following will occur in this circuit?

   A. The lights will be brighter than normal.
   B. The fuse will blow immediately.
   C. The light output will be reduced.
   D. The lights will remain on when switched off.

76. A technician is diagnosing a starting problem on a loader. He measures the voltage drop across the solenoid battery contacts. Which of the following reading would be considered maximum values?

   A. 0.5 V.
   B. 0.9 V.
   C. 2.0 V.
   D. 3.0 V.

77. A technician is diagnosing a short in an alternator charging circuit on an excavator. Which of the following test instruments would he use to diagnose this problem?

    A. Test light.
    B. Carbon pile tester.
    C. Induction tester.
    D. Ohmmeter.

78. A technician is performing a load test with a carbon pile tester to measure alternator output on a bulldozer charging circuit. What should the minimum rpm of the alternator be for this test?

    A. 1,000 rpm.
    B. 2,000 rpm.
    C. 3,000 rpm.
    D. 4,000 rpm.

79. A technician is diagnosing a fault code in the cooling temperature circuit of a diesel engine. Which of the following signal would he be testing for?

    A. An inverted digital signal with a scope.
    B. An alternating binary signal with a digital multimeter (DMM).
    C. A pulse-width modulated signal with a scope.
    D. A variable voltage signal with a voltmeter.

80. A technician is going to test a diode for proper operation. Which of the following test procedures could he use?

    A. Apply voltage across the diode and measure current flow.
    B. Apply current and measure voltage flow across the diode.
    C. Measure resistance in both directions to determine condition.
    D. Measure current output to determine the condition of the diode.

81. A technician is replacing an ECM (electronic control module) on a diesel engine. Which of the following information must be reprogrammed into the ECM?

    A. Reprogram engine horsepower rating.
    B. Reprogram customer data.
    C. Reprogram electronic unit injectors.
    D. Reprogram Programmable Read Only Memory (PROM) chip parameters.

82. A technician is going to perform diagnostic testing on a electronically controlled diesel engine with unit injectors, which involves capturing multiple frames of data over a given time. The test is triggered by the technician at an appropriate time so it can be analyzed later. Which of the following diagnostic tests will he use?

    A. Cylinder cutout test.
    B. Injector response time.
    C. Diagnostic boot test.
    D. Snapshot test.

83. A technician is testing the performance of an analog wheel speed sensor on a haulage truck. Which of the following would identify the operating principles of the sensor?

    A. Produces a DC voltage signal.
    B. Measures a percentage of ON/OFF time.
    C. Produces an AC frequency signal.
    D. Produces a variable DC voltage signal.

84. A technician is testing an negative temperature coefficient (NTC)–type temperature sensor for proper operation. Which of the following describes the operation of this type of sensor?

    A. Resistance increases as temperature increases.
    B. Resistance decreases as temperature decreases.
    C. Resistance remains the same at all temperatures.
    D. Resistance decreases as temperature increases.

85. A technician is performing a voltage drop test on the ground side of a diesel engine starter and records a voltage drop of 1.3 V. Which of the following conditions should be checked out first?

    A. This is within specifications. Nothing needs to be done.
    B. Check for high resistance at the ground connections.
    C. Check for a short in the rotor windings causing the voltage drop.
    D. Check for high resistance at the starter solenoid.

86. A technician is setting up to perform a voltage drop test on a truck with a 12 V starting system. What range must he set his digital multimeter (DMM) V-DC to?

    A. 0 to 0.1 V-DC.
    B. 0 to 2 V-DC.
    C. 0 to 20 V-DC.
    D. 0 to 30 V-DC.

87. A technician is testing the condition of a battery by using a carbon pile tester to load the battery which will allow him to determine the condition of the battery. Which of the following will occur when he turns the control knob on the carbon pile tester clockwise?

    A. The voltage and current readings increase.
    B. The resistance is increased with an increase in current flow.
    C. The resistance is decreased with an increase in current flow.
    D. The voltage will increase with a decrease in current flow.

88. A technician is checking the pulse width on a diesel engine equipped with full authority mechanically electronic unit injector (EUI)–equipped engine. Which of the following describes what is actually being measured during this test?

    A. The length of time an injector is turned ON, measured in crankshaft degrees.
    B. The length of time an injector is turned ON, measured in camshaft degrees.
    C. The ON time in percent compared to the OFF time in a given time and speed.
    D. The OFF time in degrees compared to the ON time in degrees at a given speed.

89. A technician is about to test the condition of the engine wiring that leads to a temperature sensor with a DMM. What precautions must be observed when performing this test?

    A. Use engine-specific test prods to prevent damaging connectors.
    B. Use any approved test connectors to prevent damaging connectors.
    C. Set the meter to the highest scale available when beginning.
    D. Close connectors before testing for continuity when beginning.

Block G. Environmental Control Systems (6 Questions)

90. A technician is performing a visual inspection on a haulage truck's air conditioning system. He discovers that the condenser is blocked with an obstruction. Which of the following symptoms would likely be an indication of this problem?

    A. Low airflow from the vents in the cab.
    B. The cooling system would shut down.
    C. Insufficient cooling temperatures.
    D. A noticeable loss of refrigerant would occur.

91. A technician is performing a pressure check on an excavator's air conditioning system. He observes that the low-pressure side has a reading of 9 psi with the high side at 82 psi with both readings showing on the low side. The ambient air temperature at the time of the test is at 85°F (29°C). Which of the following could be responsible for these readings?

    A. A restriction on the high side.
    B. Reading indicates normal operation.
    C. The compressor bypassing internally.
    D. Readings indicate a low refrigerant level.

92. A technician is performing a refrigerant recovery on a loader's AC system before changing an AC component. How much refrigerant can be transferred to a recovery cylinder before it is considered full?

    A. 60%.
    B. 65%.
    C. 75%.
    D. 80%.

93. A technician has performed a pressure check on the AC system of an excavator. Their findings are as follows: the low side shows higher than normal readings and the high side shows lower than normal readings. Which of the following could be responsible for these readings?

    A. Refrigerant overcharge is indicated.
    B. Refrigerant oil levels are below normal.
    C. Internal compressor damage has occurred.
    D. The vapor switch is not functioning correctly.

94. A technician suspects that the refrigerant levels are low on a loader's AC system. Which of the following symptoms would have led him to this conclusion?

    A. Condenser shows signs of overheating.
    B. High-side pressure excessively high.
    C. The compressor is cycling too often.
    D. Evaporator core ices up after 10 minutes.

95. A technician has performed a diagnostic pressure check on a haulage truck's AC system. He finds the high-side pressure to be above normal with the low-side pressure showing normal to high pressure readings. Which of the following could be responsible for these readings?

   A. The bypass valve is stuck open.
   B. The thermostatic expansion valve is leaking.
   C. The air passages in the condenser are restricted.
   D. The vapor control switch is not functioning correctly.

Block H. Structural Components, Accessories and Attachments (5 Questions)

96. A technician discovers that the excavator's drive motor shaft seal has started to leak. Which of the following could cause this to occur?

   A. The case drain line may have a kink in it.
   B. The pump suction line has collapsed.
   C. The drive track is adjusted too tightly.
   D. Excessive drive motor operating pressures.

97. A technician is performing an adjustment to a grader circle wear plates. He has run out of adjustment. Which of the following would be the considered acceptable to restore further adjustment to the grader circle?

   A. Remove shims between the wear plates and the circle.
   B. Some wear plates can be reversed to extend adjustment.
   C. The circle wear plates can be rebuilt by welding on them.
   D. Installing longer adjustment bolts will extend the life of the wear plates.

98. A technician is going to perform a track tension adjustment to a bulldozer in the field. Typically how much track sag should there be between the idler and the first carrier roller?

   A. 1/2" per foot of track length.
   B. 1.5" – 2" inches of sag.
   C. 2" – 3" inches of sag.
   D. No track sag is allowed.

99. A technician has found that the chain on the sprocket of a bulldozer is trying to climb the sprocket teeth constantly. Which of the following could cause this to occur?

    A. Excessive wear to the drive sprocket teeth.
    B. The track tension is adjusted too tightly.
    C. Excessive track misalignment.
    D. Excessive wear to the carrier rollers flanges.

100. A technician has inspected a track on a crawler tractor and found that the pin boss areas of each link shows signs of excessive wear. Which of the following could be responsible for this wear?

    A. The front idler is misaligned.
    B. The track rollers are excessively worn.
    C. Excessive drive sprocket wear.
    D. Excessive wear to the track rollers.

Practice Exam 2 ➤ 4

The following practice exam has 100 questions and follows the weighting of the Red Seal exam. The answer explanations can be found on page 83.

Block A. Common Occupation Skills (9 Questions)

1. During a normal work day in the shop, you may have been exposed to a higher than normal level of carbon monoxide (CO). Which symptom would you likely experience?

    A. A bad headache and a feeling of tiredness.
    B. Break out in a skin rash on your arms.
    C. A decrease in your blood pressure.
    D. Develop a dry cough.

2. A technician is welding a crack in the fender of a loader and inadvertently uses a rod with a broken coating on the welding electrode. What will occur to your weld?

    A. Hard spots.
    B. Poor penetration.
    C. Slag inclusions.
    D. Excessive fusion.

3. Before beginning an extensive cutting procedure on a loader bucket, a technician is going to check the contents of an acetylene bottle to see if it is full. How much pressure should be in the bottle if it is full at a shop temperature of 68°F (20°C)?

   A. 150 psi (1,034 KPa).
   B. 250 psi (1,725 KPa). ✓
   C. 315 psi (2,175 KPa).
   D. 400 psi (2,758 KPa).

4. A technician has completed a metal inert gas (MIG) weld process and discovers a high level of porosity in the weld. What may have caused this porosity in the weld?

   A. Too high of a travel speed
   B. Too low of a travel speed.
   C. Not enough shielding gas. ✓
   D. Not enough bead watering.

5. Where should you place the ground clamp when performing metal inert gas (MIG) welding on an articulating loader?

   A. The farthest distance from the weld as possible
   B. It can be located anywhere on the frame.
   C. Must be as close to the weld as possible. ✓
   D. No greater than 6'(1.8288 m) from the weld on the frame.

6. A technician is installing a new dump cylinder on a loader. Where should the load centre of gravity be located when connecting slings to raise this heavy load?

   A. Must be below the sling attachment point.
   B. Always above the sling attachment point.
   C. Never parallel with the attachment point.
   D. Always vertical to the attachment point.

7. A technician is going to lift a heavy load using a synthetic fibre sling. Which configuration will have the highest-rated lifting capacity?

   A. A single choker hitch.
   B. A single vertical hitch.
   C. A single basket hitch.
   D. A double choker hitch.

8. A technician has replaced a U-joint on a driveline and has been instructed to use an anaerobic sealant on the U-joint bolts. Which of the following identifies a characteristic of an anaerobic sealant?

   A. Remains flexible at all temperatures.
   B. Seals against lubricating oils only.
   C. Hardens in the absence of air.
   D. Hardens in the presence of air.

9. A technician has removed a cylinder head and has discovered some of the head bolts have become necked out. What would likely cause this to occur to some of the bolts?

   A. Some of the bolts were under-torqued.
   B. Some of the bolts were stretched.
   C. Dry torque was used on some of the bolts.
   D. The bolts were torqued by degrees.

Block B. Engines and Engine Support Systems (17 Questions)

10. A technician has found high silicone levels in the engine oil found during a routine oil analysis on a diesel engine. Which of the following could be responsible for this condition?

    A. Above normal air filter restriction.
    B. Damaged to the air filter housing.
    C. Damage to the turbo turbine.
    D. Above normal oil temperature.

11. During a routine maintenance inspection a technician finds the engine oil in a diesel engine crankcase has a milky, cloudy appearance. Which of the following would likely cause this?

    A. Fuel present in the crankcase.
    B. Silica present in the crankcase.
    C. Engine coolant present in the oil.
    D. Excessively high oil temperatures.

12. A technician has replaced cylinder liners in a diesel engine and notices that there is a significant amount of pitting on the major thrust side of the liners. Which of the following could cause this to occur?

    A. Excessive engine lugging.
    B. Combustion gas leakage.
    C. Lack of coolant in the radiator.
    D. Vapor bubble collapse.

13. A technician has diagnosed that a diesel engine in a loader produces light black smoke only under load. Which of the following should be performed first when diagnosing this condition?

    A. Injection timing check.
    B. Plugged particulate trap test.
    C. Fuel chemistry analysis.
    D. Air filter restriction test.

14. A technician pulls a water in fuel (WIF) alert code on a haulage truck during a routine service. Which of the following should be performed?

    A. Drain the water separator sump.
    B. Replace all the fuel filters.
    C. Replace the water separator.
    D. Add water emulsifier to the fuel tank.

15. A technician is in the process of overhauling a diesel engine cylinder head and is about to measure valve guide wear. Which of the following must be observed?

    A. The valve should be at operating temperature.
    B. The valve and guide should be at operating temperature.
    C. The valve and guide should be at room temperature.
    D. The valve should be installed into the guide for accuracy.

16. A technician is testing an engine fuel temperature sensor on an excavator with a digital multimeter (DMM) for proper operation. What type of signal does this type of sensor provide the electronic control module (ECM)?

    A. Pulse width data.
    B. Variable voltage data.
    C. Temperature data.
    D. AC voltage data.

17. A technician is going to re-program the customer data parameters with an electronic service tool (EST). Which part of the electronic control module (ECM) is this information written to?

    A. RAM (Random Access Memory).
    B. ROM (Read Only Memory).
    C. PROM (Programmable Read Only Memory).
    D. EEPROM (Electrically Erasable Programmable Read Only Memory).

18. A technician is replacing an analog throttle position sensor on an electronically controlled diesel engine in an excavator. Which of the following types are used in this type of sensor?

    A. Capacitive type.
    B. Potentiometer.
    C. Reluctor type.
    D. Stepper motor.

19. A technician has replaced a sensor that signals the actual rail pressure to the electronic control module (ECM) on a common rail engine. Which of the following components has he replaced?

    A. Rail pressure sensor.
    B. Rail pressure temperature sensor.
    C. Rail pressure hall effect sensor
    D. Rail flow limiter valve.

20. A technician is about to diagnose a low-power complaint on a loader's diesel engine by performing a leakdown test. Which of the following will be determined?

    A. Whether the rings and valves are sealing properly.
    B. Whether the rings and bearings are sealing properly.
    C. Whether the exhaust valves are in good condition.
    D. Whether the fuel pressure is within specifications.

21. A technician checks the radiator pressure on a loader and sees a sharp increase in pressure when the engine is first started. Which of the following would likely be responsible for this?

    A. Seized thermostat in the closed position.
    B. Stuck thermostat in the open position.
    C. A restriction in the radiator core.
    D. Possible cracked cylinder head.

22. A technician performing a weekly service on a loader has to determine the coolant condition in the radiator for the correct level of supplemental coolant additives and pH levels. Which of the following would indicate correct pH levels?

    A. Acidic.
    B. Neutral.
    C. Alkaline.
    D. Corrosive.

23. A technician is performing a service on a line haulage truck engine and discovers that the engine oil is very thin and extremely black. Which of the following is likely responsible for this condition?

    A. Extreme cylinder and ring wear.
    B. Bearing material contamination.
    C. Engine coolant contamination.
    D. Diesel fuel contamination.

24. A technician discovers a diesel engine in a haulage truck is producing excessive blue smoke under heavy loads. Which of the following conditions would likely be the cause of this?

    A. Turbocharger seal failure.
    B. Excessive over-fueling under load.
    C. An excessively high intake restriction.
    D. Coolant leakage into the cylinders.

25. A technician is performing an electronic unit injector (EUI) cut-out test on a diesel engine. What will occur to the injector pulse width signal when the EUI cut-out test is performed on a cylinder that is not firing?

    A. Pulse width changes; rpm decreases.
    B. Pulse width does not change; rpm increases.
    C. Pulse width changes; rpm remains the same.
    D. Pulse width does not change.

26. A technician performs a cylinder compression test on a diesel engine that has received a low-power complaint. After the engine has been warmed up to operating temperature for the test, which of the following compression values would indicate normal compression pressures?

    A. 17.3 to 20.8 Bar (250 to 300 psi).
    B. 24.0 to 29.5 Bar (350 to 425 psi).
    C. 31.0 to 38.0 Bar (450 to 550 psi).
    D. 41.5 to 55.2 Bar (600 to 800 psi).

Block C: Hydraulic, Hydrostatic and Pneumatic Systems (20 Questions)

27. A technician is going to recharge the accumulator on a loader. Why is nitrogen the only gas that must be used for this purpose?

    A. It prevents leakage after the pre-charge from the accumulator.
    B. Nitrogen is an inert gas that will not support combustion.
    C. It prevents thermodynamic pressure changes in the accumulator.
    D. It prevents piston seal binding and galling in the accumulator.

28. A technician is testing a loader with an open centre hydraulic system for proper operation. What will occur to the pump outlet pressure when the control valve is placed in the neutral position?

    A. Hydraulic pressure increases immediately.
    B. Low pressure will increase slightly.
    C. Oil flows to the tank, pressure decreases.
    D. Actuator return pressure increases.

29. A technician is reviewing the hydraulic schematic from a bulldozer to locate the source of problem with system pressure. What does this symbol represent on the schematic?

    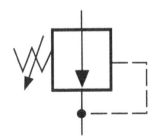

    A. Pressure relief.
    B. Pressure reducing.
    C. Priority flow.
    D. Sequence.

30. A technician is testing a loader with a closed-centre pressure-compensated hydraulic circuit that utilizes a poppet-type directional control valve. What will occur to the flow when it is placed in the neutral position?

    A. Full oil flow returning to the pump inlet.
    B. Neutral bias flow to reservoir.
    C. Varies depending on pressure.
    D. No flow is permitted to reservoir.

31. A technician is diagnosing a performance problem in a hydraulic circuit that has experienced internal wear in the hydraulic motor. Which one of the following would result from the increase in wear?

    A. Output torque increases.
    B. Output torque decreases.
    C. Output flow increases.
    D. Output flow remains constant.

32. A technician is installing a temperature sensor on the hydraulic tank. Which of the following sensors would be used for this purpose?

    A. Negative temperature coefficient (NTC thermistor).
    B. A hall effect sensor.
    C. A potentiometer.
    D. Peizo-resistive sensor.

33. A technician has diagnosed a hydraulic pump on a loader that has reduced oil flow output. Which of the following could be responsible for this symptom?

    A. Back pressure.
    B. Internal wear.
    C. Pump seal leakage.
    D. Relief valve setting.

34. A technician diagnoses a hydraulic performance problem in which a gradual slowdown of the loader's hydraulic functions occurs as the oil reaches normal operating temperature. Which of the following is likely to cause this to occur?

    A. Restricted pump outlet.
    B. Restricted pump inlet.
    C. Internal pump wear.
    D. Faulty relief valve.

35. A technician must perform one of the following steps before removing any hydraulic hoses on a hydraulic circuit safely. Which one of the following should be performed first?

    A. Release all pressure in the system.
    B. Drain the reservoir completely.
    C. Remove the reservoir cap.
    D. Have a replacement hose available.

36. A technician is about to adjust the main relief valve on a loader's hydraulic circuit. Which of the following pressures will he actually be adjusting?

    A. Pilot pressure.
    B. Cracking pressure.
    C. Port relief pressure.
    D. System relief pressure.

37. A technician is installing a new load check valve on a loader's hydraulic system. Which of the following will this installation enable?

    A. Allows oil to fill a cylinder when the pump supply is inadequate.
    B. Limits pressure peaks when the system relief valve sticks.
    C. Stop a cylinder from drifting just before it starts to move.
    D. Controls oil flow in a cylinder during pressure spikes.

38. A technician is testing the operation of a pressure sequence valve on a bulldozer. What function does this valve control in the circuit?

    A. Limits the system pressure of the primary circuit.
    B. Unloads the secondary circuit system pressure.
    C. Limits the system pressure of the hydraulic circuit.
    D. Unloads the secondary circuit oil flow.

39. A technician is going to adjust the main pressure-regulating valve. Where will the excess oil flow be routed to?

    A. Power beyond.
    B. Reservoir.
    C. Actuator.
    D. Sequence valve.

40. A technician is diagnosing a performance problem on a hydraulic system of a loader. The main system pressure only shows a maximum of 1,200 psi and it should read 1,600 psi. Which of the following could be causing this condition?

    A. Weak spring in the relief valve.
    B. A bent spool valve in the control valve.
    C. A bent cylinder rod in the hoist cylinder.
    D. Plugged strainer in the hydraulic tank.

41. A technician must flush the hydraulic system on a loader after a catastrophic failure of a hydraulic pump. Which of the following is the correct procedure to flush a hydraulic system?

    A. Use a flushing fluid with cleaning additives.
    B. Use a flushing fluid before refilling.
    C. Use the same hydraulic fluid used for the system.
    D. Use a varsol-based cleaning agent before refilling system.

42. A technician is about to remove an accumulator from a haulage truck. Which of the following must be done first to ensure safe removal of the accumulator?

    A. Bleed off any remaining hydraulic pressure from below the piston.
    B. Bleed the hydraulic pressure first, and then bleed off the nitrogen.
    C. Relieve system pressure from the hydraulic tank first.
    D. Bleed off any nitrogen gas pressure from under the piston.

43. A technician is inspecting the hydraulic system and notices that the vent on the reservoir is plugged. What effect will a plugged vent have on system performance?

    A. Pressure buildup in the tank becomes high.
    B. Increased restriction at the pump inlet.
    C. Can lead to emulsification problems.
    D. Can lead to hydraulic pump overheating.

44. A technician is replacing a defective piston pump on a loader. What should be performed first before the equipment is started for the first time?

    A. Prime the piston pump before initial start-up.
    B. Adjust the engine speed to specifications.
    C. Back off the main pressure relief before start-up.
    D. Replace the inlet hose before start-up.

45. A technician is testing the inlet side of the charge pump of a hydrostatic drive circuit for proper operation with a manometer that is calibrated in inches of mercury. What reading would be considered acceptable?

    A. 5″
    B. 10″
    C. 15″
    D. 20″

46. A technician is going to replace a hydraulic hose on a loader which has number 6 marking on the hose. Which of the following would this number refer to?

    A. 6 bar burst pressure.
    B. 3/8″ of an inch inside diameter.
    C. 3/8″ of an inch outside diameter.
    D. 6 bar working pressure.

Block D. Drivetrain Systems (14 Questions)

47. A technician is diagnosing a performance problem on a loader with a countershaft power-shift transmission with a remote-mounted torque converter. During a torque stall test, the technician notices that the high-speed driveline between the torque converter and transmission is creeping in a certain gear range. Which of the following would likely be responsible for this symptom?

    A. Excessive clutch disc wear.
    B. Torque pressure regulating spring defective.
    C. High backpressure in torque converter drain line.
    D. High backpressure in converter drain line.

48. A technician is diagnosing a countershaft power-shift transmission overheating problem. Which of the following could likely cause the transmission oil to overheat?

    A. A clutch pack is dragging or slipping.
    B. High oil level in the transmission sump.
    C. High clutch pressure in the transmission.
    D. Excessive high-torque inlet pressure.

49. During a routine service inspection, a technician observes that the transmission clutch pressure is lower in one gear range than in any of the other gear ranges. Which of the following could likely cause this?

   A. Excessively worn clutch piston rings or internal clutch drum wear.
   B. Excessively weak or broken clutch piston return springs.
   C. Defective clutch pressure regulating valve spring.
   D. Excessive internal wear to impeller and turbine blades.

50. A technician is checking the condition of a drive chain on a grader. Which of the following specifications would be correct to determine maximum accepted chain elongation dimensions?

   A. 50% of original length.
   B. 20% of original length.
   C. 3% of original length.
   D. 6% of original length.

51. A technician has diagnosed a torque converter high stall speed condition. While reading the diagnostic procedure in the shop manual, the technician comes across the term "vortex flow." Which of the following describes this term correctly?

   A. Flow of oil around the circumference of the torque converter.
   B. Spiral flow of oil between the impeller, turbine, and stator.
   C. Movement of oil created by centrifugal force of the impeller.
   D. Flow of oil during the coupling phase of operation.

52. A technician has diagnosed a power-shift transmission output shaft seal leak on a loader. Which of the following should be checked first?

   A. Check lubricant level and top up.
   B. Remove seal and check wear surface.
   C. Check if transmission breather is plugged.
   D. Pressure-test the transmission housing.

53. A technician is installing the stator with a "spragg clutch" on a torque converter during an overhaul of the transmission. Which way should the stator freewheel in after it is in place?

   A. In the direction of flywheel rotation.
   B. In the opposite direction to crankshaft rotation.
   C. Either direction. It does not matter.
   D. Must be locked to the stator support sleeve.

54. A technician has diagnosed a yoke bore misalignment on an articulating loader's mid-ship driveline. Which of the following would lead to this misalignment condition?

    A. Over-torquing of the yoke retaining nut.
    B. Excessive driveline torque during operation
    C. Universal joint retaining bolts overtorqued.
    D. Improper universal joint lubrication intervals.

55. A technician is going to replace the centre bearing on a loader's mid-ship driveline assembly. Which of the following should be performed during the replacement procedure?

    A. Measure the shims during removal of the old centre bearing.
    B. Remove the bearing seals on the new bearing and pack with grease.
    C. Record the driveline angles before removing the old bearing.
    D. Reuse the old mount assembly and just replace the bearing.

56. A technician is diagnosing a performance problem with a loader's power-shift transmission circuit. When the technician checks the engine's high idle rpm, it is within specs, but the converter stall speed is 200 rpm above specification. Which of the following could be responsible for the low torque stall?

    A. The torque converter's stator has excessive wear.
    B. The torque converter's turbine bearing is damaged.
    C. The torque converter has worn turbine blades.
    D. The torque converter has low charge pressure.

57. A technician is performing an overhaul on a differential. The pattern on the teeth of the ring gear shows up as top heel contact on the gear tooth. Which one of the following must be done to try and correct this condition?

    A. Move the pinion away from the gear.
    B. Increase the mounting distance.
    C. Move ring gear away from pinion.
    D. Move the pinion towards the gear.

58. A technician has diagnosed a bevel pinion failure in a differential. Which of the following conditions would have contributed to this result?

    A. When both wheels are under heavy torque load in reverse.
    B. When one wheel is spun faster than the other at high speed.
    C. When one of the axle bearings preloaded is set too loose.
    D. When the bearing preload is excessive on the pinion gear.

59. A technician is overhauling a differential and is about to reset the excessive backlash reading on the gear set. Which of the following must be done to correct this condition?

    A. Move the pinion gear towards the ring gear.
    B. Move the ring gear away from the pinion.
    C. Move the ring gear towards the pinion.
    D. Move the pinion gear away from the ring gear.

60. A technician is overhauling the forward clutch pack on a loader's countershaft power shift transmission. The technician discovers that the clutch discs have been slipping and are worn excessively. Which of the following must be replaced along with the clutch discs?

    A. The transmission pump must be replaced as well as the forward drum.
    B. The clutch pressure regulating valve must be replaced.
    C. The piston rings on the piston and support sleeve have to be replaced.
    D. The forward drum and the clutch discs must be replaced as a set.

Block E. Steering, Suspension, Brake Systems, Wheel Assemblies and Undercarriage (14 questions)

61. A technician is adjusting the steering geometry so that the haulage truck has better directional stability on a roadway. Which of the following would be adjusted to achieve this?

    A. Adjust the toe-in by a few degrees.
    B. Adjust the camber setting by a few degrees.
    C. Adjust the steering axis inclination.
    D. Adjust the toe out by a few degrees.

62. A technician is going to charge a suspension cylinder on a 200-tonne haulage truck with nitrogen gas. Which of the following must be done before he or she begins?

    A. Compensate for temperature.
    B. Recover the existing nitrogen.
    C. Drain nitrogen from adjacent cylinder.
    D. Load vehicle to rated capacity.

63. A technician is diagnosing a performance problem with the emergency brake automatic brake application process on a loader. What should result when the brake pressure drops below the accumulator pressure switch limit?

    A. Service brakes will apply.
    B. Increased pump cycle time.
    C. Park brakes will apply.
    D. Emergency brakes will apply.

64. A technician is diagnosing a complaint that the brake charge valve on a loader is cycling too often. Which of the following must be checked for first?

    A. Service brake application pressure too high.
    B. Pilot circuit pressure too high.
    C. Low accumulator pre-charge pressure.
    D. Oil leak to the gas side of the accumulator.

65. A technician is checking the service brake application pressure on a loader with spring-applied hydraulically released brakes. Which component regulates the service brake application pressure?

    A. Pilot sequence valve.
    B. Accumulators.
    C. Charge valve.
    D. Brake foot valve.

66. A technician is about to change a hydraulic brake hose on the wheel end of loader with spring-applied hydraulically released brakes in the applied position. Which of the following must be performed before this work can proceed safely?

    A. Release the accumulator gas pressure with the brakes applied.
    B. None. Wheel ends have no pressure with the brakes applied.
    C. Release the accumulator hydraulic brake pressure first.
    D. Release the emergency brake pressure first.

67. A technician is performing a routine oil change on a rear axle of a loader equipped with an internal hydraulically applied wet disc brake system. He discovers that the differential is overfilled with two gallons of oil. Which of the following could be responsible for this to occur?

    A. Brake system lubrication oil is over-filling the differential due to seal failure.
    B. High pressure brake piston seal is leaking oil to the differential housing.
    C. Charge pressure too high, causing the piston seal to leak into the differential.
    D. The brake cooling pressure is set too high, causing seal leakage.

68. A technician is testing the accumulator pressure switch on a loader with an internal wet disc brake system. He places a jumper wire across the contacts on the accumulator pressure switch. Which of the following should occur?

    A. Turns on low brake pressure indicator and applies the emergency brake.
    B. Turns off low brake pressure warning light and releases emergency brakes.
    C. Turns on the emergency brake application impending light in the cab.
    D. Applies the emergency brakes due to high accumulator brake pressure.

69. A technician is testing a haulage truck with hydraulic drum brakes and finds that when he depresses the brake pedal it slowly creeps down to the floor. There are no external leaks. Which of the following could be responsible for this symptom?

    A. The brake shoes need to be re-adjusted.
    B. Internal bypassing of the master cylinder piston seals.
    C. The brake shoes are worn beyond safe limits.
    D. Wheel cylinders are bypassing internally.

70. A technician is inspecting a track on a bulldozer which has a damaged link counterbore. Which of the following can cause this type of wear to occur?

    A. Seal failure.
    B. Loose track.
    C. Sprocket wear.
    D. Recoil spring damage.

71. A technician has diagnosed a track on a dozer with excessive pin and bushing wear. Which of the following procedures would need to be performed?

    A. Build up the pins and bushings with weld.
    B. Replace the track links and bushings.
    C. Rotate the pins and bushings half a turn.
    D. Track is operating too loosely. Tighten up the track.

72. A technician is performing routine maintenance on a dozer track by checking the track for signs of wear. There are signs of wear on one side only of an idler on one track only. Which of the following could be responsible for this symptom?

    A. Damaged recoil springs.
    B. Track misalignment.
    C. Track tension adjustment too tight.
    D. Excessive wear to the sprocket.

73. A technician is inspecting the rim of a large haulage truck that is in for a tire change and notices that the one of the rim components has developed a small crack. Which of the following is an acceptable repair procedure?

    A. Replace the rim components. Welding is not allowed.
    B. Only certified welders are allowed to weld on rim components.
    C. Rims can only be welded on if hard resurfacing rods are used.
    D. Rims must be preheated before welding can occur properly.

74. A technician is about to replace a tire on a wheeled dozer and has to inject liquid ballast into the tire. Which of the following would be the correct amount of ballast for this application?

    A. 20% by volume.
    B. 40% by volume.
    C. 75% by volume.
    D. 90% by volume.

Block F. Electrical and Vehicle Management Systems (15 Questions)

75. The operator has reported that the charging system indicator light is remaining on after the loader has started. Which of the following would identify the problem?

    A. The alternator is not charging the battery.
    B. The charge voltage is excessive.
    C. The battery has an internal short.
    D. The battery is overcharging.

76. After a technician diagnosed a slow cranking speed problem and replaced the starter on a loader, it still has a slow cranking speed. Which of the following could cause this slow cranking speed condition?

    A. Starter bendix drive binding.
    B. Battery connectors corroded.
    C. Defective solenoid hold-in winding.
    D. Defective solenoid pull-in winding.

77. A technician is testing a voltage regulator on a loader for proper operation. Where would he connect the meter leads to determine the correct operation of the voltage regulator?

    A. Terminal "F" to ground on the alternator.
    B. In series with the alternator output terminal.
    C. (BAT) terminal to alternator ground.
    D. Terminal "F" to the alternator output terminal.

78. A technician is testing the output of an alternator with a carbon pile tester. Which of the following readings would indicate a failed rectifier diode bridge?

    A. The output will be reduced by 2/3.
    B. The output will be reduced by 1/3.
    C. The output will be not be affected.
    D. High voltage output will result.

79. A technician is diagnosing a code for high engine oil temperature on a log skidder. He is about to test the oil temperature sensor to determine if it functioning correctly. Which test would the technician perform initially to determine if the sensor was actually working?

    A. Use an electronic service tool (EST) to determine the sensor's voltage range.
    B. Use a scope to determine the sensor's wave form.
    C. Test the sensors magnetic properties on metal.
    D. Heat the sensor and check for resistance changes.

80. A technician conducted an alternator output test using a carbon pile tester on an excavator that has two batteries in parallel with low current output. Which of the following should be checked first to determine the cause?

    A. The voltage drop normal on the ground from the alternator to the block.
    B. An excessive voltage drop at output terminal of the alternator.
    C. The two batteries internal resistance is low.
    D. Too many accessories turned on at the same time.

81. A technician has performed a charging circuit voltage drop test on a loader with a 12-V alternator; which of the following would be considered maximum for voltage drop reading?

    A. 0.1 V.
    B. 0.2 V.
    C. 0.5 V.
    D. 1.0 V.

82. A technician has performed a diagnostic test on an analog throttle position sensor. In one of the tests he measures the voltage at the sensors output terminal while changing the position of the throttle position sensor. What reading would be produced from low idle to the high idle position?

    A. The voltage would remain the same.
    B. The voltage would vary proportionately.
    C. The AC voltage would decrease by 100% overall.
    D. The voltage percentage of ON time versus OFF time would vary.

83. A technician has diagnosed a fault that is currently present in a circuit of an electronically controlled diesel engine. Which of the following would indicate the status of the fault?

    A. Current.
    B. Inactive.
    C. Active.
    D. Historical.

84. A technician has diagnosed an electrical short to positive in the headlight circuit of a loader. Which of the following could occur from this condition?

    A. Circuit resistance decreases.
    B. Voltage drop for the circuit increases.
    C. Magnetic interference increases.
    D. Voltage induction increases.

85. A technician has performed a voltage drop test on a loader's starting circuit. Which of the following must occur for this test to produce accurate results?

    A. The battery open circuit voltage must be at 13.2 V.
    B. There must be current flow in the circuit being tested.
    C. There must be no current flowing in the circuit being tested.
    D. The voltmeter must be placed in series in the circuit.

86. A technician is using a digital multi-meter (DMM) with it set to DC volts to test the voltage drop in an electrical lighting circuit on a loader. How must the voltmeter be connected in the circuit to measure voltage drop?

    A. Parallel to the circuit.
    B. In series in the circuit.
    C. In series parallel to the circuit.
    D. In phase with the circuit.

87. A technician is testing a light relay for a suspected internal short in the coil. He is using a digital multi-meter (DMM); what reading should be expected when testing the coil resistance?

    A. Infinite reading on the current scale.
    B. Higher than specified resistance.
    C. Lower than specified resistance.
    D. Infinite reading on the ohms scale.

88. A technician is going to test the electronic control module (ECM) output voltage reference voltage (V-ref) to a coolant temperature sensor circuit. Which of the following voltage is found in this circuit?

    A. 2.5 V.
    B. 5 V.
    C. 8 V.
    D. 12 V.

89. A technician is testing the operation of an analog turbo boost sensor on an electronically controlled diesel engine. Which of the following sensor designs is used in this type of turbo boost sensor?

    A. Piezo-resistive.
    B. Pulse generator.
    C. NTC thermistor.
    D. Potentiometer.

Block G. Environmental Control Systems (6 Questions)

90. A technician is diagnosing an air conditioning operational noise complaint. Whenever the A/C is turned on, a rumbling noise seems to be coming from the A/C clutch area. Which of the following is the most likely cause of this noise?

    A. The refrigerant level is below normal.
    B. Internal compressor bearings are damaged.
    C. Compressor clutch bearing failure.
    D. The drive belt needs to be replaced.

91. A technician is diagnosing an A/C performance problem. He installs a set of test gauges and finds that the high side pressure is well above normal with a below normal reading on the low side. Which of the following could be responsible for this condition?

    A. Lower than normal air duct pressure.
    B. Indicates a restriction in the orifice tube.
    C. The compressor clutch fuse is blown.
    D. Low refrigerant levels in the system.

92. A technician is diagnosing a performance complaint on a fixed orifice tube A/C system. A visual inspection reveals that the compressor clutch is kicking in and out several times a minute and the evaporator outlet line is warm to the touch. He also notes that there is no frost on any of the A/C components. Which of the following could be responsible for this?

    A. The evaporator is flooded.
    B. A low refrigerant charge.
    C. A restricted accumulator.
    D. High refrigerant levels in the system.

93. A technician is checking the operation of an orifice tube type A/C system on a haulage truck. What state would the refrigerant be in at the outlet of the receiver dryer on a properly functioning A/C system?

    A. High pressure gas.
    B. Low pressure liquid.
    C. Low pressure gas.
    D. High pressure liquid.

94. A technician has diagnosed a problem in an excavator's A/C system and has found a thermal expansion valve's sensing bulb with a broken line. Which of the following symptoms would occur with this condition present?

   A. Duct outlet temperature increases.
   B. Duct outlet temperature decreases.
   C. Low side pressure increases.
   D. High side pressure increases.

95. A technician has diagnosed a performance problem in an A/C system in a haulage truck. The outlet duct air temperature is higher than normal when the AC is operating. Which of the following could be responsible for this symptom?

   A. The heater control valve is stuck closed.
   B. Debris has collected in the evaporator core inlet.
   C. The heater control valve is stuck open.
   D. A vacuum leak at the heater control valve.

Block H. Structural Components, Accessories and Attachments (5 Questions)

96. A technician is attempting to diagnose a complaint from an operator of a boom truck. Whenever he operates the power take off (PTO) he hears a loud whining noise coming from the PTO area. Which of the following could be responsible for this noise?

   A. Too much backlash between the mating gears.
   B. A broken shift fork in the PTO shifter.
   C. Too little backlash between the mating gears.
   D. A broken PTO gear tooth on a mating gear.

97. A technician is diagnosing a hydraulic winch performance problem on a log skidder. When the operator engages the winch, he finds that it operates slower than normal and is not moving at its rated capacity. Which of the following could be responsible for this symptom?

   A. The hydraulic winch motor may have an internal leak.
   B. An incorrectly adjusted control valve spool.
   C. The restricted case drain on the winch motor.
   D. The main relief valve pressure is too high.

98. A technician is performing routine maintenance to a grader circle. He finds that the ball joint knuckle on the circle has become loose. Which of the following should be performed to correct this condition?

    A. Replace the ball joint with a new one.
    B. Remove some shims to tighten.
    C. Add some shims to tighten.
    D. User thicker viscosity grease.

99. A technician is changing grouser shoes on a track loader. Which procedure must be followed to install the pads correctly on the track?

    A. Install the grouser shoes with the wide part over the link.
    B. Install the grouser shoes with the narrow part over the link.
    C. Install the grouser shoes with shims under the plow bolts.
    D. Install the grouser shoes with grease between pad and link.

100. A technician has to order a new wire rope to replace the one on a winch of a skidder. Which of the following wire construction classifications should be used for this application?

    A. 6 x 19.
    B. 7 x 21.
    C. 4 x 12.
    D. 6 x 30.

Answers and Explanations for Practice Test 1 ➤ 5

Block A. Common Occupation Skills (9 Questions)

1. **Answer C is correct.**
   None of answers A, B or D have any official meaning in the Workplace Material Hazardous Material Information System (WHMIS) document. A Material Safety Data Sheet (MSDS) contains information on the potential hazards (health, fire, reactivity, and environmental) and how to work safely with a specific product. It also contains information on the use, storage, handling and emergency procedures related to the hazards of a particular material.

2. **Answer D is correct.**
   The only acceptable repair that should be performed before any welding takes place is to replace the cable with the insulation cracks. None of A, B, or C are recommended.

3. **Answer C is correct.**
   If the flywheel housing to crankshaft concentricity is outside of the specification, in some cases it can be re-aligned. Begin by removing the dowels and replace with oversize locating dowels if they are available. Loosen the flywheel mounting bolts and then snug them up by hand so they are able to move when struck with a rubber mallet. Based on the reading of the dial indicator adjust the position of the flywheel housing by tapping with the rubber mallet to achieve the best run-out possible. Repeat the adjustment procedure until the readings meet the required specifications.

4. **Answer B is correct.**
   The pitch of the screw thread on a standard micrometer spindle is 40 threads per inch (TPI). One revolution of the thimble advances the spindle face toward or

away from the anvil face precisely 0.025". The reading line on the sleeve is divided into 40 equal parts by vertical lines that correspond to the number of threads on the spindle.

5. **Answer B is correct.**
Only a soap-and-water mixture is recommended for use on an oxy-fuel system for leak-testing purposes. Response A or B are not safe or effective. Response D would be dangerous to do as it could lead to a fire or explosion. No oil of any kind should ever be used on an oxy-fuel system for leak-testing purposes.

6. **Answer B is correct.**
It's important to get the correct shade tint when gas welding or brazing. Use the appropriate face shields, spectacles, or goggles, depending on the particular job, to protect faces and eyes from welding hazards. Reponses A, C, or D are incorrect because each lens shade number is too light or too dark.

7. **Answer D is correct.**
In general, a safety factor of 5:1 is generally used for wire rope. Certain sling fittings such as hooks will deform beyond usefulness before breaking and cannot be assigned a definite numerical design factor.

8. **Answer A is correct.**
The open side of the seal contains a spring that keeps the lip in contact with the crankshaft keeping the oil in and the dirt out. Using any kind of lubricant on the outside of the seal to the housing side may cause the seal to work its way out. A light coating of oil on the seal face is all that is required before installation. The lip of the seal does not need any extra lubrication as the oil from the engine side of the housing will provide enough oil to keep the seal lubricated.

9. **Answer D is correct.**
A torque-to-yield fastener is a bolt that is torqued to the state of tension, causing it to become elongated slightly. It is not recommended to reuse this type of fastener. It is often used in high-load situations that have a higher risk of fatigue-related failure, like bolts that are used on cylinder heads or connecting rods often are torque-to-yield bolts.

Block B. Engines and Engine Support Systems (17 Questions)

10. **Answer C is correct.**
    Many radiator caps have a vacuum valve built into the cap that allows air or coolant to re-enter the radiator as the coolant cools, preventing radiator hose collapse.

11. **Answer A is correct.**
    A radiator with no pressure in it will cause the coolant to boil sooner as the boiling point of the coolant is higher when the radiator has pressure in it. One psi of pressure increase raises the boiling point of coolant by approximately (1.65°C) 3°F.

12. **Answer A is correct.**
    Torsional stress causes excessive twisting to occur in the crankshaft which can lead to cracks developing in the main journal cheeks. Misalignment of the cylinder would not cause any torsional vibration to occur in the crankshaft. Crankshaft lubrication failure would cause bearing damage to the crankshaft. Excessively worn rod bearings do not cause torsional vibration but results in lower oil pressure and eventually bearing failure.

13. **Answer B is correct.**
    A manometer calibrated in inches of water is used to measure the negative low-pressure restriction on the suction side of the air filter. When the restriction reaches 25″ the filter is generally replaced. It's important to note that the maximum restriction occurs under maximum engine loads, not at high engine idle speed.

14. **Answer A is correct.**
    Only internal nozzle wear will cause excess fuel to leak past the nozzle and result in high back leakage. Answer B, C, and D will not affect back leakage. Excessive back leakage is caused by wear to the nozzle and body of the nozzle assembly. The only thing that seals this area is the tight running clearances which are in the order of 2 to 5 microns, depending on the nozzle type. This allows for some back leakage to occur for lubrication purposes.

15. **Answer A is correct.**
    The shorter line will result in a quicker pressure rise due to a lower volume of fuel in that line. The pumping capacity of the plunger hasn't changed; therefore, it will reach cracking pressure slightly faster, causing a slight advance to the timing for

that cylinder. Response B and C would not be correct as explained above. Response C is not correct as cracking pressure is controlled by the spring pressure in the injector, not by line length. Response D is not correct as the length of line will not increase cracking pressure. The cracking pressure is controlled by the tension of the nozzle spring which is adjustable. The more spring tension is applied the higher the cracking pressure. The fuel pressure must overcome the spring pressure before the fuel can be injected into the combustion chamber.

16. **Answer D is correct.**
The opacity meter is the only instrument that measures particulate using the light extinction principle. Response A, B, and C do not measure opacity but are used to measure various exhaust gases such as CO (carbon monoxide) and NO_2 (nitrogen dioxide). VOC (volatile organic compound) analyzers are not generally used on diesel engines, but are often used to detect concentrations of heavy hydrocarbons in gas or liquid media.

17. **Answer D is correct.**
Lower-than-normal operating temperatures would result in incomplete combustion which would elevate the hydrocarbon levels in the exhaust. Oxides of nitrogen and nitrous oxides usually become elevated when combustion chamber temperatures are too high. Carbon dioxide is present in combustion and will not change significantly as a result of temperature changes in the engine.

18. **Answer B is correct.**
Carbon monoxide (CO) is converted to carbon dioxide (CO_2) in a catalytic converter. A catalytic converter has no effect on nitrogen or NO_x. The catalytic converter is not considered a filter and does not filter out particulate or hydrocarbon. That is the job for a diesel particulate filter (DPF).

19. **Answer A is correct.**
Isolating the vehicle's electrical system from the equipment chassis during welding will prevent any chance of welding current from passing through the equipment electrical system which could lead to electrical or electronic component damage. Disconnecting only one of the battery leads could still cause a problem with transient welding voltages damaging the equipment. Disconnecting the computer connections only could still cause electrical components to be damaged by stray voltages if the ground path from welding went through any electrical components.

20. **Answer A is correct.**
    The aneroid valve limits the rate of fuel delivery during acceleration with spring pressure opposing boost pressure, ensuring there is enough air to burn the fuel completely, minimizing the amount of black smoke in the exhaust.

21. **Answer A is correct.**
    Fuel filters that have a 2 micron rating will not allow water to pass through and will eventually plug up and stop the engine from running. Fuel pressure is regulated with a relief valve and will not increase past a certain point, protecting the filter from bursting. Water does not get absorbed into the primary fuel circuit. The fuel filter media is built to withstand fuel-system pressures and will not collapse.

22. **Answer C is correct.**
    Worn piston rings and cylinder liners will allow air to enter the crankcase and escape out of the dipstick tube. If the intake valve were not sealing properly, air would escape into the intake manifold. If the exhaust valve was leaking, the air would escape out the exhaust. Valve guide wear cannot be determined with a cylinder leakdown test.

23. **Answer A is correct.**
    Excessive oil leakage indicates that there is excess clearance between the rod journal and the connecting rod bearings. A leakdown test would not show excessive end play to the crankshaft, excessive bearing crush or insufficient rod journal side clearance.

24. **Answer B is correct.**
    Once the turbocharger is replaced, pour some clean oil into the turbocharger oil inlet and turn the compressor wheel by hand to spread oil on the bearings. Before starting the engine, disable the electronic control module (ECM) by pulling the fuse that sends power to the ECM and crank the engine over until oil pressure is achieved—usually around 5 to 10 seconds.

25. **Answer A is correct.**
    The boiling point of the coolant will increase by 1.66°C or 3°F for every 1 psi (0.07 Bar) of additional pressure in a radiator.

26. **Answer A is correct.**
    To determine the efficiency of an air-to-air aftercooler, the air temperature drop between the inlet and outlet of the cooler must be compared to the specifications

provided by the engine manufacturer. The main purpose of the air-to-air intercooler is to lower the air temperature of the compressed air. Response B, C or D do not address the temperature difference across the intercooler.

Block C. Hydraulic, Hydrostatic and Pneumatic Systems (20 Questions)

27. **Answer B is correct.**
Normal wear over time causes the pump gears to wear the thrust plates on the sides of the pump housing. These are often replaceable on many types of gear pumps. An increase in end play of the gears is a sign that the pump needs replacement or needs to be rebuilt. Neither increased backpressure nor a higher operating rpm would cause an appreciable increase in pump wear. Dirt is another factor that will wear out a pump quicker than normal. An increase in pump pressure will not affect pump wear greatly, although wear was present in the pump; the amount of bypassing oil would increase with a higher operating pressure.

28. **Answer D is correct.**
When a priority flow is required to a designated circuit before feeding a secondary circuit, a priority compensated flow control can be used to accurately split the flow between the circuits. When it is required, a set restriction provides the priority side with a set flow rate to the circuit. The remaining oil is directed to the secondary circuit. As flow rate changes, the secondary circuit will receive more or less flow while the priority side remains satisfied.

29. **Answer C is correct.**
The correct gas pre-charge pressure must be maintained for proper functioning and optimum service life for a gas-charged bladder-type accumulator. In a bladder-type accumulator, excessive pre-charge can drive the bladder into the poppet assembly during discharge, causing damage to the poppet assembly and/or the bladder.

30. **Answer C is correct.**
A single-acting telescoping hydraulic cylinder is not self-bleeding so that any air trapped in the cylinder must be removed. The port relief valve, control valve and hydraulic pump do not need to be bled as they are self-bleeding in normal operation, allowing oil and air to flow through each component.

31. **Answer D is correct.**
    The oil must be up to its normal operating temperature to ensure a correct reading. In cold temperature climates, this is even more important as the oil viscosity thickens when the temperature is below freezing. If there was internal wear in the hydraulic pump, checking it at below operating temperatures might conceal internal wear due to the thicker oil at below operating temperatures.

32. **Answer C is correct.**
    Control valves generally have extremely tight clearances between the spool and body. If the problem is intermittent, a quick check of the spool travel is in order. If the spool does not self-centre itself each time it is returned to neutral, it may be sticking, causing the load to drift down on occasion.

33. **Answer C is correct.**
    Two types of seals are generally used in double-acting cylinders, single bi-directional along with the more common type, uni-directional seal. If the packing is bypassing, extend the cylinder to maximum travel in one direction, remove the hose at the stem end, and load the cylinder with pressure on the opposite end. No oil or very little oil should leak at the end with the hose removed. If there is no evidence of bypassing, repeat the test at the other end of the cylinder. If the seal is defective, a leak will be evident at one end or the other.

34. **Answer A is correct.**
    Suction-line leaks or leaks between the tank and the pump can introduce air into the intake of the pump. Air can also leak in at the shaft seal of the pump that allows air to leak in to the pump. Once air bubbles form, they stay locked in the oil and don't come out of the oil in the reservoir quickly. Cavitation occurs when air bubbles in the suction-line fluid are subsequently imploded in the pump when the oil is pressurized, causing pitting to the thrust faces.

35. **Answer D is correct.**
    If the cycle time in a closed centre system increases substantially, an internal leak is the only response that could likely make this condition occur. A low-relief valve setting, internal pump wear, or a restricted pump inlet would not increase the cycling time, but may affect speeds. Only a pressure loss internally could cause the system to cycle more frequently.

36. **Answer A is correct.**
    Low oil levels would cause air to be pulled into the pump causing a jerky response because of air's compressibility. As the air is compressed, the actuator slows down.

Once the air is compressed sufficiently to overcome system resistance, it then moves quickly again, causing the jerky movement.

37. **Answer A is correct.**
Any time a component such as a main relief is replaced, it is good practice to back off the relief valve setting before starting the equipment back up. This will prevent any unexpected pressure from occurring. Once the equipment is running, the pressure can be increased gradually to the required setting.

38. **Answer C is correct.**
The swash plate angle on the variable displacement motor can be adjusted to slow down the output speed as it is variable in design. If it were a fixed displacement design, you would not be able to adjust its output as the stroke is not adjustable.

39. **Answer C is correct.**
If the hydraulic tank was over-filled, the oil could escape out of the vent due to the agitation in the tank during operation. A plugged hydraulic suction filter would cause pump starvation but not cause oil to leak out of the vent on the tank. The main relief valve sticking would not have any effect on the oil in the tank. Excessive pump speed would not cause the tank to spill oil from the vent, but could cause maximum pump pressure to be exceeded and also cause seal failure on the suction side.

40. **Answer C is correct.**
Excessive cycling could occur as a result of less gas charge in the accumulator, which would result in quicker dissipation of the oil pressure on the hydraulic side of the accumulator. As pressure is used up in the hydraulic system, the gas charge helps maintain pressure longer in the hydraulic circuit as the gas pressure drops at a slower rate than the hydraulic pressure, which drops rapidly on its own.

41. **Answer A is correct.-**
Air would be pulled into the suction side of the pump which would cause the oil to foam. Any air that enters the pump gets compressed as it exits the pump, causing a high-pitched whining sound and eventually leading to pump failure.

42. **Answer C is correct.**
A high case drain could indicate an internal leak, which could lead to a higher frequency of cycling depending on the rate of case drain. The higher the case's drain, the higher the cycling frequency rate.

43. **Answer C is correct.**
    Any time the oil level is low in the hydraulic tank air would be pulled into the pump. As the bubbles collapse during pressurization of the oil, cavitation occurs causing a loud whining noise and eventual pump damage.

44. **Answer B is correct.**
    Most manufacturers recommend a maximum restriction at the pump inlet of 5″ of mercury. Measurement in inches of water is not used to measure hydraulic pump restriction. One inch of mercury is equivalent to 13.608″ of water. A common reading usually found in exhaust system restriction testing is 29″ of H_2O. Twenty-nine inches of Hg is normally associated with AC system testing.

45. **Answer A is correct.**
    Any time a seal is leaking internally, the volume of oil flow would be reduced, causing the boom to operate at a slower rate. The lowering cycle would not be affected sufficiently to notice a difference in speed. Erratic boom operation is usually caused by air in the oil or possibly a bent stem. The seal bypassing would not cause increased load-lifting capability.

46. **Answer A is correct.**
    If the pressure drops when the cylinder is at full stroke, the problem is likely the fault of the port relief. It does not come into play when the pressure is checked with the cylinder at rest on the stops. A bent rod would likely have no effect on the pressure, and a hydraulic restrainer restriction would not have any effect on system pressure.

Block D. Drivetrain Systems (14 Questions)

47. **Answer B is correct.**
    The low converter-in pressure indicates that the torque converter pressure regulating valve is not working correctly, as it should not release excess oil to the lube circuit until converter pressure is normal.

48. **Answer A is correct.**
    The engine must be at maximum rpm with the loader in gear and the brakes on and the turbine stalled. The rpm indicated during this test shows the general condition of the power train.

49. **Answer A is correct.**
    The side gears only turn when one wheel is turning faster than the other wheel, as when the truck is turning a corner. If the side gears had excessive wear, it would make noise whenever one wheel was turning faster than the other. Heavy toe contact, chipped ring gear teeth or excessive backlash would be responsible for this intermittent symptom, but would likely be noisy while driving in a straight line as well.

50. **Answer B is correct.**
    Unequal tire size would cause the bevel gears to rotate steadily during straight line operation. One wheel would rotate farther than the other due to unequal tire size causing the bevel gears to rotate the same way they do during cornering. These gears are not designed to rotate all the time and should only come into play when the wheel speeds are not equal.

51. **Answer B is correct.**
    During operation the differential can experience heavy loading. The thrust screw is adjusted to help eliminate excessive ring gear deflection to maintain proper tooth contact between the ring gear and pinion.

52. **Answer C is correct.**
    The fact that there is no clutch slippage in reverse 2nd indicates that the 2nd gear clutch is not responsible for the slippage as it only occurs in forward 2nd. This indicates that forward clutch-pack discs have excessive wear and need to be replaced.

53. **Answer B is correct.**
    The transmission oil cooler circuit should generally be the first place to perform a pressure differential test across the inlet and outlet of the cooler. This test indicates whether or not there is a restriction in the cooler circuit or low flow, which would indicate an oil flow problem which could lead to an overheat problem.

54. **Answer B is correct.**
    The technician should check the clutch pressure variation by noting the pressure differences between clutch pressures while shifting the transmission through its gear ranges. Any significant variations would indicate excessive clutch leakage. For example, if the clutch pressure was 200 psi in forward first and forward second and 190 psi in reverse first and reverse second, it's easy to see that there is a problem with the reverse clutch bypassing internally.

55. **Answer B is correct.**
The transmission cooler circuit is the easiest test to perform quickly by checking the temperature drop across the cooler with an infrared heat gun during high-load conditions. All the other responses could play a factor but are more difficult to check. When troubleshooting always eliminate the easy checks first.

56. **Answer A is correct.**
The accepted method to measure the pitch of chain is to measure the centre-to-centre distance across one link. Measuring across more than one link helps determine the wear on the chain.

57. **Answer A is correct.**
Whenever abnormal wear appears on one gear in a set, the ring and pinion should always be replaced as a set, not individually. All ring and pinion sets are manufactured as a matched pair. The ring and pinion gears should never be mixed with gears from another set, as they have been carefully machined together for optimum contact. The ring gear and the pinion gear will have matching "hand etched" numbers engraved on them to verify they are a set.

58. **Answer B is correct.**
The term "backlash" refers to the clearance between the pinion teeth and the crown gear teeth. The setting is accomplished by either moving the pinion in or out, or by moving the crown gear left or right. This movement changes the clearance between the pinion and the crown gear teeth and also affects the tooth contact pattern.

59. **Answer B is correct.**
The clutch pressure regulating valve is responsible for regulating the clutch pressure during engagement. If it is sticking it will cause initial gear engagement to be slow. Low clutch pressure will not affect shifting. Low converter-out pressure indicates the converter safety relief valve is stuck in the transmission cover. The lube pressure regulating valve would have no effect on shifting.

60. **Answer C is correct.**
Low lube pressure would indicate an oil supply problem, as the lube circuit is the last circuit to receive oil in the transmission circuit from the pump. It generally does not have a regulating valve and relies on excess supply to meet the circuit's oil requirements.

Block E. Steering, Suspension, Brake Systems, Wheel End and Undercarriage (14 Questions)

61. **Answer A is correct.**
    Adding or subtracting shims increases or decreases the vertical clearance. Steering knuckle vertical clearance should be checked and adjusted usually during a rebuild. Clearance between the knuckle and the top of the axle eyelet should be adjusted to specification so as to provide a minimum clearance of 0.020″ (0.508 mm) between the knuckle and the bottom of the axle eyelet.

62. **Response A is correct.**
    If the toe setting were incorrectly set, tire scuffing could occur. The scuffing action produced the featheredge wear pattern across the tread face. Feathering is a condition wherein the edge of each tread rib develops a slightly rounded edge on one side and a sharp edge on the other. Running your hand over the tire thread, you can usually feel the sharper edges before you'll be able to see them. If toe-in is set correctly and this wear pattern still occurs, it is usually due to worn bushings in the front suspension, causing the wheel alignment to shift as the vehicle moves while it is driven.

63. **Answer A is correct.**
    Worn suspension parts would cause the toe adjustment to change erratically during operation, causing wear to both sides of the tires. Under-inflated tires can produce the same type of wear.

64. **Answer C is correct.**
    The strut is fastened to the cylinder rod on one side and the frame of the truck on the other side. The king pin is not part of the rear suspension. The torque rod and equalizer beam are part of the spring suspension.

65. **Answer C is correct.**
    All the air must be purged from the accumulator before charging with nitrogen. Air left in the accumulator would pose an explosion hazard if hot oil entered the gas side.

66. **Answer C is correct.**
    Straight kingpins require a draw key to locate them in the correct position in the steering knuckle. Draw keys must be installed, one from each side of the axle. Do not install both pins from the same side of the axle.

67. **Answer B is correct.**
    Suspension cylinders charge pressure should not vary by more than 50 psi (345 KPa) on most truck suspension cylinders. A gas charge pressure of 100 psi (690

KPa) is too great of a variation between cylinders. The struts should never extend fully when the box is empty.

68. **Answer B is correct.**
    The knuckle pin bushings have excessive wear that allows the spindle to move inboard toward the frame rail. This could result in a negative camber angle. Excessive wear in the trucks rear spring shackles and pins does not affect camber. A seized bearing between the steering knuckle and the axle eye won't affect camber and neither will shimming between the spring and axle.

69. **Answer B is correct.**
    The spring-applied brake has a series of high-pressure springs that apply the brakes. High-pressure oil is required to hold back the springs during disassembly. Therefore the spring force must be mechanically caged during the removal of the friction discs. Once the springs are safely caged, the discs can be removed safely.

70. **Answer A is correct.**
    Only the cut-out pressures are actually adjustable on a charge valve. The cut-in pressure goes up as a result of adjusting the cut-out pressure as it is controlled by the same spring.

71. **Answer B is correct.**
    Once the minimum pressure is reached, the charge valve will open the circuit to the accumulators and allow the brake pump to charge the accumulator circuit until it reaches cut-out pressure.

72. **Answer A is correct.**
    The brakes on this type of system are released hydraulically and any restriction in the foot brake valve return circuit would cause the brakes to apply slowly as pressure release would be impeded by a restriction.

73. **Answer C is correct.**
    The shuttle valve function is to provide two separate paths for brake application, one from the service brake pedal and one from the emergency brake circuit. Either one will apply the brake wheel ends.

74. **Answer B is correct.**
    The only possibility for reduced pressure from the choices offered would be restricted brake pedal travel often caused by debris under the brake pedal interfering with its full motion of travel.

Block F. Electrical and Vehicle Management Systems (15 Questions)

75. **Answer D is correct.**
    The lights are getting back feed from another circuit wire in the harness, keeping them on. Back feeding occurs when a power wire insulation covering wears through from vibration or melts when excessive current comes in contact with another power wire in a harness, causing the power to flow, bypassing the light switch.

76. **Answer A is correct.**
    The accepted maximum voltage drop across a starting solenoid is 0.5 V. Any reading above this is considered to be excessive and the solenoid should be replaced.

77. **Answer D is correct.**
    The ohmmeter is the tool to use to find a short in a circuit by measuring the resistance of each circuit and comparing it to known values. None of the other instruments are suitable for measuring resistance in a circuit.

78. **Answer B is correct.**
    In order for this test to be valid, a minimum rpm recommended for the alternator must be achieved to get an accurate reading of the output. The alternator's charging output increases in proportion to the electrical load on the charging system and alternator speed. Maximum output is generally achieved at alternator speeds of at least 2,000 rpm.

79. **Answer D is correct.**
    Thermistors are normally used to measure temperature on cooling systems. The thermistor changes resistance with temperature changes and will vary the voltage signal sent back to the electronic control module (ECM).

80. **Answer C is correct.**
    The resistance mode typically is used only if a multi-meter is not equipped with a diode test mode. When performing resistance testing, the diode should be taken out of the circuit since it can produce a false reading while in the circuit. The more accurate way to test a diode is by measuring the voltage drop across the diode. The forward-biased resistance of a good diode should range from 1,000 Ω to 10 MΩ. And the reverse-biased resistance of a good diode displays over limit (OL).

81. **Answer B is correct.**
    The replacement electronic control module (ECM) does not come programmed with customer data and must be reprogrammed in to the ECM before operation. Customer data includes such things as road speed limits, idle shutdown timer, and cruise control parameters. A password is required to reprogram these parameters.

82. **Answer D is correct.**
    The snapshot test is designed to capture data that the electronic control module (ECM) is tracking. The snapshot can be triggered by a code or be manually triggered by the operator while an event or symptom is occurring. This data can be saved or printed out for later analysis.

83. **Answer C is correct.**
    When the sensor passes by the sensor ring tooth, it generates an AC voltage signal that is sent to the electronic control module (ECM) indicating a position or speed. The ECM is capable of counting the number of occurrences in a given time to indicate speed accurately.

84. **Answer D is correct.**
    There are two types of thermistor that are used to measure temperature on mobile equipment. Depending on temperature range and manufacturer, with the negative temperature coefficient design the resistance decreases with a corresponding increase in temperature. The positive coefficient design is the opposite, with resistance increasing as the temperature increases.

85. **Answer B is correct.**
    The carbon pile load tester is generally used to generate current flow. This is much preferable to using the starter to turn over the engine. The carbon pile provides a steady load, something that cranking the engine won't always give you. This results in more accurate test results than you would get when using the starter to load the circuit.

86. **Answer B is correct.**
    A voltage setting from zero to 2 V-DC should be sufficient to measure voltage drops which usually sit around the 0.2–0.5 V range, occasionally going higher.

87. **Answer B is correct.**
    The carbon pile tester has been the tool of choice for many years as far as battery testing goes. The tester applies a load to the battery that can simulate the load of a starter during engine cranking conditions. The established testing protocol dictates that 50% of the cold cranking amps (CCA) be applied as a load to the

battery for 15 seconds. During this test, as the current is applied and the voltage is monitored it must stay above 9.6 V during the test in an ambient temperature of 50°F or 10°C or higher.

88. **Answer A is correct.**
The injector cut-out test measures the duration in crankshaft degrees that the injector is actually injecting fuel into the engine. This test is used to determine which cylinder is not contributing its portion of the load. The test cuts out one injector at a time at a given rpm and records the change in fueling for the rest of the injectors. There should be a significant change in response time when one injector is disabled compared to the average response time when all injectors are firing normally.

89. **Answer A is correct.**
Any time that injector harness connectors are to be probed with test tools, it's imperative that the correct probes are used. The use of incorrect probes will result in loose connections after they are re-assembled, resulting in intermittent problems that are created by a loose connection or changes in resistance across connectors.

Block G. Environmental Control Systems (6 Questions)

90. **Answer C is correct.**
When foreign material blocks airflow through the condenser, it prevents the entire refrigerant from condensing back to a liquid, which causes a drop in cooling efficiency. Freeze-over of the evaporator is generally caused when the temperature is too low with the moisture in the air that passes through it freezing on the core. Poor airflow from the dash vents may be caused by problems with the air distribution system. A loss of refrigerant would likely be caused by a leak in the system.

91. **Answer D is correct.**
This problem is generally a result of a low refrigerant charge. A malfunctioning compressor would not cause low pressure on both sides. It generally results in low pressure on the high side and high pressure on the low side. A restriction on the high side would give you high pressure on both sides.

92. **Answer D is correct.**
A refrigerant cylinder should never be filled past the 80% level. A safe percentage of room must be left for expansion that could occur due to temperature fluctuations that could occur to the cylinder during storage.

93. **Answer C is correct.**
    During normal operation, the compressor lowers the low-side pressure and raises the high-side. A damaged compressor cannot maintain this pressure condition in the two sides. When this occurs, cooling will diminish, especially at low engine speeds. During higher engine speeds the pressures may rise closer to normal reading but can't sustain these values at low engine speeds.

94. **Answer C is correct.**
    Excessive cycling could indicate a low charge level. The AC pressures depend on the type of refrigerant and the ambient temperature when the pressure is checked to determine if the system is properly charged. For example, at a ambient temperature of 85°F (29°C), R134A would register a static pressure around 90 psi of pressure on both gauges. A lower static reading suggests the system is low on charge. The leak must be found and repaired before repairs can be considered complete.

95. **Answer C is correct.**
    Poor air flow across the condenser is likely the only choice of the four responses that could cause high-side pressure to be above normal with the low-side pressure showing normal to high pressure readings. The three other responses would cause lower than normal pressure on the low side.

Block H. Structural Components, Accessories and Attachments (5 Questions)

96. **Answer A is correct.**
    If the case drain becomes restricted, the pressure could build up too high in the motor case and cause the seal to leak. The seal is not designed to hold any pressure; its purpose is to keep dirt out and case drain oil in.

97. **Answer A is correct.**
    Play in the circle, lift cylinders, and moldboard slide should be checked daily. No free play is desirable. It is permissible to reverse the wear plates and shim accordingly to establish correct dimensions.

98. **Answer B is correct.**
    The track is properly adjusted when you have 1.5″–2″ of sag between the idler and first carrier roller on a conventional track on most dozers.

99. **Answer A is correct.**

If the drive sprocket teeth have excessive wear, the track would tend to try and climb the sprocket teeth. In some designs you are able to reverse the drive sprockets as they are bolted on in three or more sections; otherwise they should be replaced as this will cause excessive stress to the track.

100. **Answer A is correct.**

The front idler keeps the track chain in alignment while the dozer is traveling and also acts as a shock absorber and track tensioner. Any misalignment will cause excessive misalignment wear to the pin boss areas of each link.

Answers and Explanations for Practice Test 2   6

Block A. Common Occupational Skills (9 questions)

1. **Answer A is correct.**
   Carbon monoxide (CO) gas is one of the most widespread and dangerous industrial hazards. It is a common cause of on-the-job gas poisoning, sometimes leading to death. It can be lethal at concentrations as low as 1,000 parts per million (ppm). The carbon monoxide builds in the blood, which prevents the normal amount of oxygen from being absorbed into the blood stream causing headaches and a feeling of tiredness. For example, the time-weighted limit average for exposure to carbon monoxide in Ontario is 25 ppm (parts per million).

2. **Answer C is correct.**
   The coating on the rod has a layer of a chemical which acts as a cleaning and purifying agent. It protects the weld pool and solid metal from atmospheric contamination and helps in removing impurities from the weld pool. The impurities float on the surface of the pool and are easily removed after it cools down.

3. **Answer B is correct.**
   Any change in ambient temperature affects the pressure in an acetylene cylinder at a much faster rate than it affects the pressure in an oxygen cylinder. Pressure in an oxygen cylinder will go up or down only about 4% for each 20°F (7°C) change in temperature above 70°F (20°C). An acetylene cylinder with a pressure of 250 psi at 70°F (1,725 KPa at 20°C) will have a pressure of 315 psi at 90°F (2,175 KPa at 31°C) and at a temperature of 50°F the pressure would drop to 190 psi at (1,300 KPa (9°C). The ambient temperature has to be taken into account when estimating how much acetylene a cylinder contains.

4. **Answer C is correct.**
Hydrogen can be unintentionally introduced during the welding operation through contaminants within the welding area. Exposure of the molten weld metal to the surrounding atmosphere during the welding process is one consideration when examining a porosity problem. This situation may occur as a result of inadequate gas shielding during welding.

5. **Answer C is correct.**
The arc generated by welding needs a smooth flow of electricity though the equipment to complete the electrical circuit. I use this analogy as an example to explain the importance of a good ground. If you were to crimp a garden hose slightly, you would have noticed that the flow at the end of the hose would be reduced. Electricity and fluids respond to restrictions in much the same way. Technicians sometimes make the mistake of attaching the work clamp (electrical ground) to a painted or a rusty surface. Both of these surfaces can act as electrical insulators and do not allow the welding current to flow easily, resulting in unstable current flow. A good indication of a poor connection is a ground clamp that is hot to the touch. Welding current will seek the path of least resistance just as water does, so if care is not taken to place the welding ground close to the arc, the welding current may find a path unknown to the technician and possibly damage electrical components on the equipment.

6. **Answer A is correct.**
The centre of gravity is the point around which an object's weight is evenly balanced. A suspended object will always move until its centre of gravity is directly below its suspension point. The load's centre of gravity must be accurately estimated and located directly under the lifting eye and below the lowest sling attachment point before lifting the load. If an object is symmetrical in shape, its centre of gravity will lie at its geometric centre. The centre of gravity of an irregularly shaped object may be difficult to locate.

7. **Answer C is correct.**
This type of configuration is used to support a load by attaching one end of the sling to the hook, then passing the other end under the load and attaching it to the hook. Always make sure that the load does not turn or slide along the rope during a lift and that the centre of gravity remains below the sling. Be aware that the capacity of the basket hitches is also affected by their sling angles.

8. **Answer C is correct.**
   The characteristic of anaerobic sealants is that they remain liquid until the oxygen is removed. When an anaerobic adhesive is sealed between a nut and a bolt on a threaded assembly, it hardens to form a tough cross-linked plastic that adheres to many metals.

9. **Answer B is correct.**
   Over-torquing will neck out the head bolts. When over-stretched they actually have a reduced diameter in the area where the metal has been stretched. This reduced area has experienced deformation, and will not hold the same clamping force if tightened to specification again. This will in fact cause the bolt to stretch more (loosing tension) and can eventually lead to a fractured bolt.

Block B. Engines and Engine Support Systems (17 Questions)

10. **Answer B is correct.**
    The damaged air filter housing may not completely seal the intake system, allowing dirt to enter the intake where it gets deposited on the cylinder liner walls that are covered with oil. The dirt works down into the crankcase by piston action contaminating the oil.

11. **Answer C is correct.**
    Small amounts of coolant can give oil a caramel color while a larger quantity will give it a milky appearance. An oil analysis will confirm the presence of coolant in the oil. The need for corrective action is immediate as this could have a negative effect on crank bearings, as it inhibits the effectiveness of the lubricant.

12. **Answer D is correct.**
    During engine operation the piston changes direction as it travels up and down. Side thrust is imparted by the rods as the power changes from the linear up-and-down motion of the piston to the rotary motion of the crankshaft. During the power stroke, the side of the liner in contact moves first toward the coolant and then away from it rapidly many times per second, setting up a vibration in the liner. This movement causes vapour bubbles to form as the liner moves away from the coolant, causing them to collapse or implode as the liner moves back. This is known as cavitation erosion.

13. **Answer D is correct.**
    A water manometer gauge can be used to check the restriction on the suction side of the air filter housing. The test must be conducted under maximum load as this is the only time the reading will be at its highest level. The maximum reading should not exceed 25 " of water on the manometer under maximum load.

14. **Answer A is correct.**
    The first thing that must be done is to drain the water separator. Diesel fuel will always contain a certain percentage of water. The goal is to keep water levels within acceptable limits, well below the saturation point. Most equipment manufacturers specify that no water must reach the engine. In reality it's hard to keep water out. Saturation points in diesel fuel vary from roughly 50 ppm to 1,800 ppm depending on the ambient temperature.

15. **Answer C is correct.**
    As with any measurements performed on engine components, they should be performed with the components at or near room temperature. Any increase in temperature above room temperature will give you an inaccurate reading as compared with the specifications, as metal expands when heated and shrinks when cooled.

16. **Answer B is correct.**
    Thermistors are widely used on engines to provide variable voltage information to the electronic control module (ECM) for coolant, fuel, engine oil, and intake air temperatures as well as any other process that requires temperature monitoring. The ECM translates the voltage signal into a temperature and displays it on a dash gauge, scanner, or laptop when requested.

17. **Answer D is correct.**
    The electronically erasable programmable read only memory (EEPROM) contains customer data that is programmable and proprietary data for customer's application that can be modified using a certain tools ranging from an EST (Electronic Service Tool) to software on a laptop. EEPROM provides the electronic control module (ECM) the ability to write-to-self. This area also allows operational fault codes as well as other data to be written to this area for future reference.

18. **Answer B is correct.**
    A typical potentiometer is a three-wire variable resistor. It has a reference voltage, ground, and output voltage. A signal proportional to the motion of a throttle pedal is sent to the electronic control module (ECM) through the output. The

device moves a contact wiper over a variable resistor to produce a signal sent to the ECM. As the wiper moves over the variable resistor, the resistance path changes as the supply voltage is divided between the signal and ground. The ECM interprets the signal voltage into a pedal count and compares it to stored data in the ECM, adjusting the injector pulse width accordingly.

19. **Answer A is correct.**
    The *actual* rail pressure signal is transmitted to the electronic control module (ECM) by the rail pressure sensor. If the *desired* rail pressure is higher than required, the rail pressure control valve opens to lower fuel pressure in the common rail, sending the excess fuel back to the return circuit. If the actual rail pressure is lower than the *desired* rail pressure, the pressure control valve will close, permitting pressure to rise in the common rail.

20. **Answer A is correct.**
    The leakdown test is often used to determine the condition of the compression rings and intake and exhaust valves and seats. The test is done on an engine when it is not running with the cylinder at top dead centre on the compression stroke. Pressure is fed into a cylinder through a glow plug or injector hole. Leakdown tests tend to rotate the engine and require some method of holding the crankshaft in the proper position for each tested cylinder.

21. **Answer D is correct.**
    The crack in the cylinder head or a bad head gasket often allows combustion gases to flow into the cooling system during engine operation, causing the sharp rise in radiator pressure. In some cases, if the crack is large enough it will force the coolant to spill out the overflow.

22. **Answer B is correct.**
    Coolants degrade over time as the ethylene glycol breaks down into acids. This occurs more quickly in engines operating at higher temperatures or those that allow more air into cooling systems. The coolant should be tested annually, particularly where the coolant is used in severe conditions. One reliable test ensures the pH is still above 7.0. It is typically not good practice to allow coolant to operate below a pH of 7.0.

23. **Answer D is correct.**
    Fuel has extremely high detergent capabilities and will dilute the oils viscosity making it extremely thin and black looking, but also causes rapid wear to the crankshaft bearings due to its lubrication destroying properties.

24. **Answer A is correct.**
    If the turbo housing has developed a crack, or the internal seals have blown, oil will start to leak into your exhaust system. As this oil burns off, it will produce distinctive blue-grey smoke under load, which will probably become more apparent as the engine revs increase.

25. **Answer D is correct.**
    To determine if a faulty fuel injector is causing the cylinder to misfire, a cut-out test is performed on the engine. There are two types of cut-out test: manual and automatic. There will be no change in pulse width or rpm in the cylinder that is not contributing to the engine power.

26. **Answer C is correct.**
    Only answer C is within the guidelines for normal compression pressure on a typical electronically controlled diesel engine. Answer A and B are too low and D is too high.

Block C. Hydraulic, Hydrostatic and Pneumatic Systems (20 Questions)

27. **Answer B is correct.**
    Using any gas that supports combustion, including compressed air in an accumulator, could lead to an explosion, if oil gets in to the gas side of the accumulator.

28. **Answer C is correct.**
    Whenever the control valve on an open centre system is placed in neutral, the pump oil flow is diverted back to the tank. No oil is flowing into the circuit while the control valve is in neutral; it all returns to the tank.

29. **Answer B is correct.**
    The symbol represents a pressure-reducing valve. This type of valve is designed to reduce higher inlet pressure to a steady lower downstream pressure regardless of changing flow rate or varying inlet pressure. The valve is very sensitive to pressure changes and immediately reacts to maintain the desired downstream pressure.

30. **Answer B is correct.**
    This type of valve when used in a closed-centre pressure compensated hydraulic circuit will send the flow back to the reservoir whenever the directional control valve is placed in neutral.

31. **Answer B is correct.**
When internal wear takes place in a hydraulic motor and the motor experiences internal leakage as a result of excessive clearances within the motor, it will result in reduced volumetric efficiency. A major power loss in hydraulic systems is usually experienced as a result of internal leakage in motors.

32. **Answer A is correct.**
The negative temperature coefficient (NTC) thermistors internal resistance goes down when heated. When used in temperature measuring applications on equipment, the electronic control module (ECM) interprets the changing voltage signal into actual temperature readings. Hall effect sensors are generally used to determine position. Potentiometers are often used to indicate throttle position. Peizo-resistive sensors are used to measure low pressure accurately, such a turbo-boost pressure.

33. **Answer B is correct.**
Low oil flow at the pump outlet is generally associated with internal pump wear. On a gear pump the wear is generally associated with excessive thrust plate wear. On piston pumps the wear is often on the pistons or piston seals which results in higher-than-normal running clearances, which causes higher internal leakage.

34. **Answer C is correct.**
Problems associated with internal pump wear often go unnoticed until the oil is up to operating temperatures. The cold oil is thicker and will bypass less until it heats up and becomes thinner. All pumps have a certain amount of internal leakage built into them for lubrication purposes.

35. **Answer A is correct.**
Before working on a hydraulic system, it is of the utmost importance that all pressures have been dissipated before removing any hoses or components. Failure to do so will result in an unexpected release of oil which could cause personal injury or death. Note: you may still have pressure in the hydraulic tank even though you have relaxed all the system pressures. Some hydraulic tanks on certain types of equipment are pressurized and must also be vented.

36. **Answer D is correct.**
The system pressure is determined by the main relief valve setting. The pilot pressure is controlled by a pilot pressure regulating valve. The term "cracking pressure" is a term that is used to describe what occurs at the moment relief pressure on a relief valve is achieved.

37. **Answer C is correct.**
Generally when load checks are defective, they allow a cylinder to drift momentarily just before it starts to move. These types of check valves prevent pulling or pushing loads from accelerating uncontrollably during movements in the load direction.

38. **Answer A is correct.**
During operation of certain types of hydraulic circuits, there are times when two cylinders need to stroke in a planned sequence. When two cylinders are controlled by a single directional valve, the cylinder with the lowest resistance always strokes first. If the actuator with the least resistance is first in the sequence, the circuit runs smoothly without additional valving. The pressure sequence valve can be used to force fluid to take the path of greatest resistance.

39. **Answer B is correct.**
Pressure-regulating valves are found in every hydraulic system. The regulating valve keeps system pressures safely below a desired upper limit and maintaining a set pressure in part of a circuit.

40. **Answer A is correct.**
A weak spring in the relief valve could cause the maximum system pressure to become low over time. Neither a bent spool in the control valve, a bent cylinder rod, or a plugged strainer is capable of causing system pressure to become low.

41. **Answer C is correct.**
The accepted practice is to use the same hydraulic fluid that is used in the system to flush the hydraulic circuit. Using another type of product could cause compatibility issues with the original fluid or possibly damage the O-rings and seals.

42. **Answer B is correct.**
Both sides of the accumulator must be bled before removal from the truck. Dissipating the hydraulic pressure only leaves the nitrogen gas charge in the accumulator, which still poses a hazard.

43. **Answer B is correct.**
If atmospheric pressure is prevented from entering the tank freely, there is a danger of creating a partial vacuum in the tank which could affect the pump's ability to receive oil causing cavitation. Atmospheric pressure acting on the oil in the tank ensures that the pump receives an adequate volume of oil.

44. **Answer A is correct.**
Whenever new pumps are installed on a hydraulic system, it is mandatory to prime them with oil before attempting to startup the system, especially piston pumps, which can be severely damaged in seconds due to the close tolerances between the pistons and housing bores.

45. **Answer A is correct.**
The acceptable limit for pump restriction on a pump is measured in inches of mercury and should not exceed the manufacturer's recommended levels.

46. **Answer B is correct.**
Hydraulic hose numbering is given in sixteenths of an inch. A number 6 hose would be 6/16″ or 3/8″ measured on the inside diameter of the hose.

47. **Answer A is correct.**
When the high-speed driveline shows signs of creep during a torque stall test, it generally points to either clutch disc wear or low clutch pressure in that clutch combination. Remember, two clutches are actually used: a speed clutch and a directional clutch. If the clutch pressure is not low, the discs need to be replaced.

Block D. Drivetrain Systems (14 Questions)

48. **Answer A is correct.**
A slipping transmission clutch pack in a power-shift transmission with excessive wear in the clutch piston rings or excessive clutch disc wear will generate excessive heat, causing the transmission oil temperature to increase beyond the normal range.

49. **Answer A is correct.**
The observation indicates that the clutch pressure is only lower in one gear range, which eliminates the pump wear and defective clutch pressure regulating valve spring. Weak or broken return springs would not affect clutch pressure. Excessive wear to the impeller and turbine blades has no effect on the clutch pressure, but will affect the torque stall speed and performance.

50. **Answer C is correct.**
The general accepted specification for measuring chain wear for elongation is that 3% of its original length, as this would cause the chain to start climbing the sprocket, increasing sprocket wear significantly.

51. **Answer B is correct.**

    "Vortex flow" in a converter occurs any time the turbine and impeller are not turning at the same speed. The highest amount of vortex flow occurs during a converter stall test when the turbine is stopped and the impeller is turning at engine speed.

52. **Answer C is correct.**

    The first step is to verify that the transmission breather is not plugged. This would cause pressure buildup in the transmission housing which could cause the seal to leak.

53. **Answer A is correct.**

    The "spragg clutch" allows the stator to freewheel in one direction only and locks up when you try and rotate it in the opposite direction. It must be installed so it freewheels in the direction of flywheel rotation so that it can freewheel during the coupling phase of operation.

54. **Answer B is correct.**

    Distortion and misalignment to the yoke is often caused by excessive driveline torque, which causes excessive twisting torque that can cause the yoke bores to become distorted and misaligned. Use an alignment bar with the correct diameter to inspect for misalignment by placing the bar through the lug holes simultaneously. If the yoke has been distorted, it should be replaced.

55. **Answer A is correct.**

    Whenever a technician replaces a centre bearing on a driveline, the shims should be measured during removal of the old bearing to maintain the original position when the new centre bearing is installed.

56. **Answer D is correct.**

    If the converter is experiencing low charge pressure, the stall speed will increase, as the pressure in the converter contributes to the efficient operation of the converter assembly.

57. **Answer D is correct.**

    Moving the pinion towards the gear will move the pattern to a more desirable position on the tooth. An increase to the shims for the pinion carrier is called for to move the pattern to the desired location.

58. **Answer B is correct.**
Bevel pinion failures are often caused by excessive high wheel spin on one wheel of an axle as such as when the truck loses traction on one side. This causes the bevel gears to turn at high speeds, something they are not designed to do.

59. **Answer C is correct.**
Moving the ring gear towards the pinion will result in decreasing the backlash. Any changes to the backlash must be followed up with a gear pattern check to ensure a good contact pattern is maintained.

60. **Answer C is correct.**
The cause of the worn discs is likely the result of internal piston clutch pressure leakage in the forward clutch pack. This is caused by excessive piston ring wear on the clutch piston and the clutch drum support shaft.

Block E. Steering, Suspension, Brake Systems, Wheel Assemblies and Undercarriage (14 questions)

61. **Answer C is correct.**
The steering axis inclination is the angle between the centreline of the steering axis and vertical line from the centre contact area of the tire as viewed from the front. As the spindle travels in an arc, the tire/wheel assembly raises the suspension and forces the tire/wheel assembly to seek the centre return point and return to neutral when corning is completed. Because there is a tendency to maintain or seek a straight-ahead position, less positive caster is needed to maintain directional stability.

62. **Answer A is correct.**
The amount of nitrogen gas will depend on the ambient temperature the work is performed in. As with all compressed gases, the volume and pressure can change significantly with ambient temperature changes.

63. **Answer D is correct.**
The accumulator pressure switch will activate the automatic emergency brake application if the accumulator pressure drops below a set value.

64. **Answer C is correct.**
    When the accumulator gas charge becomes low due to a leak, the brake pressure will drop quicker with each service brake application, causing the charge valve to charge the accumulators more frequently than the specification stipulates.

65. **Answer D is correct.**
    Although the accumulator and charge valve are in the same circuit, only the foot brake valve actually regulates the amount of brake application pressure to the wheel end.

66. **Answer B is correct.**
    A spring applied brake has no pressure when the brakes are applied because they are applied with spring force and require pressure to release them.

67. **Answer A is correct.**
    The internal seal between the brake cooling circuit and the differential housing has developed a leak, causing the brake cooling oil to leak into the differential housing, causing the housing to be over-filled.

68. **Answer A is correct.**
    When the contact of the accumulator pressure switch is jumped, the indicator lamp in the operator's compartment illuminates and the emergency brakes are applied. This test is used in some cases to check the operation of the emergency brake circuit. Bypassing the accumulator pressure switch simulates low accumulator pressure, which should apply the emergency brakes automatically.

69. **Answer B is correct.**
    As there are no external leaks in the brake system, the only other thing that would allow the brake pedal to drift down slowly would be an internal leak in the master cylinder that allows the oil from the pressure side to slowly leak past the seal to the vented side of the master cylinder.

70. **Answer A is correct.**
    Seal failure of the link counterbore seal would quickly allow dirt to enter the area increasing the wear rate significantly. The seals are there to keep dirt out of the link counterbore area.

71. **Answer C is correct.**
    With certain types of track designs, the rotation of the pins and bushing has been an accepted method of repair to increase track life. The pins and bushing wear on one side only. By rotating those 180°, the track dimensions can be restored to acceptable limits.

72. **Answer B is correct.**
    Generally when wear occurs unevenly as in one side of a component, misalignment should be considered as a cause. In this case, wear on one side of the idler is not considered normal. Wear and misalignment could be responsible for this condition.

73. **Answer A is correct.**
    Welding is not permitted on any rim components at any time. The components must be replaced. Welding heat may cause unwanted metallurgical changes to the metal such a decreasing of the hardness of the metal, which could cause a rim component to fail.

74. **Answer C is correct.**
    Calcium chloride was a popular tire ballast as it is heavy, provides freeze protection, and is inexpensive to purchase. The negative side of calcium chloride is that it is extremely corrosive. When it comes in contact with the rim it will eat away at the metal in time. In recent years, better products in liquid ballasting have been developed without some of the major disadvantages such as corrosion acceleration. An agricultural by-product of sugar beet processing is becoming popular as a base ingredient in tire ballast. It has good freeze protection and does not cause corrosion to metal.

Block F. Electrical and Vehicle Management Systems (15 Questions)

75. **Answer A is correct.**
    The indicator light in the dash of the operator's compartment is there to warn the operator that the alternator is no longer charging the battery. When the key is turned on, power flows through the warning lamp, through the field coil, and then to ground, causing the lamp to illuminate. Once the alternator is at full output, voltage from the diode trio equals the battery voltage. At this point in time, with 12 V on both sides, the lamp goes out.

76. **Answer B is correct.**
    The starting circuit consists of large cables that carry the high starter current during engine cranking. High resistance in the cable ends caused by corrosion can greatly slow the cranking speed of the engine, especially in cold weather. Diesel engines need to turn over at 150 rpm or more in cold weather in order to generate enough compression heat to ignite the fuel charge. Cranking the engine for more than 30 seconds at a time without allowing for a cool down period can lead to starter burn-up. Discharged or worn-out batteries can also cause the same problem.

77. **Answer C is correct.**
    When testing the voltage regulator, we first check the voltage from the input pin to ground. This is to make sure that voltage is being supplied to the regulator. If the regulator isn't receiving voltage, it will not be able to output its rated regulated voltage; this is why we do this test. The voltage that we should read here should be higher than the voltage the regulator is rated to output, normally 1 to 2 V higher.

78. **Answer B is correct.**
    Alternators produce alternating current which is converted to direct current (DC) by a six diode rectifier located inside the back of the alternator. Diodes only allow current to flow in one direction: three positive diodes control the positive side of the AC sine wave, while three negative diodes control the negative side. If one of the diode bridges fails, then only one side of the wave form gets picked up and sent to the battery. Ripple voltage is another way to check for bad diodes. To measure ripple voltage, switch your digital multimeter (DMM)) to alternating current (AC) and connect the black lead to a good ground and the red lead to the battery (BAT) terminal on the back of the alternator.

79. **Answer D is correct.**
    If they apply heat to the sensor and test the internal resistance of the temperature sensor, they should get an increase or decrease in resistance as temperature changes. This would indicate that the sensor was at least functioning as intended and is not likely the problem.

80. **Answer B is correct.**
    A high voltage drop at the output terminal will affect the output of the alternator. Excessive voltage drop caused by high resistance can reduce the charging current significantly.

81. **Answer B is correct.**
    The typical accepted standard for maximum voltage drops in a charging system would be below 0.2 V. Anything above these values would lead to problems.

82. **Answer B is correct.**
    The analog throttle position sensor operates on a variable resistance principle. As the throttle moves, the resistance changes the output voltage and current.

83. **Answer C is correct.**
    Any fault that is present or actually occurring during operation is considered to be active. Once the condition clears up, the fault will go historical or inactive depending on the engine manufacturer.

84. **Answer A is correct.**
    A short to positive generally causes lower resistance in the circuit as the path for the power to flow is shortened, bypassing some of the circuit resistance.

85. **Answer B is correct.**
    Electrical voltage drop can vary and is dependent on current flow. Unless you operate the circuit with current flowing through it, you can't measure voltage drop without current flow, because an ohmmeter's battery can't supply the current that normally flows through most circuits. Ohmmeter tests usually can't detect restrictions as accurately as a voltage drop test.

86. **Answer A is correct.**
    To measure voltage drop you must connect the voltmeter across the circuit load parallel to the circuit and not in a series with the circuit. In phase with the circuit has no relevance when measuring voltage drop.

87. **Answer C is correct**.
    Any time there is a shorter path for electricity to flow, there is generally a decrease in resistance in that circuit.

88. **Answer B is correct.**
    A typical coolant thermistor-type temperature sensor will have a 5-V signal going to the sensor varying the voltage signal it sends back to the electronic control module (ECM) when the resistance changes due to coolant temperature changes. The variable voltage in the sensing line going back to the ECM is interpreted as temperature of the coolant.

89. **Answer A is correct.**
Pressure sensors come in a variety of designs. Generally a piezo-resistive type sensor is well suited to measure small changes in boost pressure quickly and accurately.

Block G. Environmental Control Systems (6 Questions)

90. **Answer B is correct.**
The low level of refrigerant will not in itself cause this type of noise. Some of the new A/C systems won't allow the compressor to engage if the low side pressure is too low, preventing damage to the compressor. Noise from internal bearing damage in the compressor will only appear when the compressor is actually turning, which only occurs when the A/C clutch is engaged. A defective A/C clutch bearing would be noisy anytime the engine is operating regardless of A/C operation. Although a drive belt can become noisy when worn, it will not cause this type of noise.

91. **Answer B is correct.**
A common cause of below normal low side pressure is a restriction in the orifice tube on a fully charged system. During operation a restriction in the orifice tube will causes the high side to become elevated and can quickly lead to compressor damage, as this can interfere with oil flow to the compressor. Low air duct pressures would point to a circulation problem. A blown fuse in the clutch circuit would prevent the compressor from turning and would not cause high pressure on the high side. Low refrigerant charge would show up as lower than normal pressure on both gauges. You would not get a high pressure reading on the high side.

92. **Answer B is correct.**
Whenever the clutch cycles too quickly and there is no frost on any of the components, it usually can be traced back to a low refrigerant charge level. Any restriction in the accumulator would lead to frosting. A flooded evaporator would lead to frosting on the evaporator outlet line as well as the compressor suction line. Because there are no signs of frosting on any of the A/C components, only low levels of refrigerant would likely cause this.

93. **Answer C is correct.**
The receiver drier traps any liquid that flows from the evaporator. A desiccant helps remove moisture and debris. Only low pressure gas should enter the compressor; any liquid would cause severe damage to the compressor. High pressure gas only flows from the compressor, into the condenser. The condenser removes the heat

from the refrigerant which changes state to a high pressure liquid. It then passes through the orifice tube, a restriction that causes a pressure drop. From there the low pressure liquid leaves the orifice and enters the evaporator core.

94. **Answer A is correct.**
    When the high pressure liquid enters the expansion valve it allows a portion of the refrigerant to enter the evaporator. In order for the higher temperature fluid to cool, the flow must be limited into the evaporator to keep the pressure low and allow expansion back into the gas phase. The thermal expansion valve has a sensing bulb connected to the suction line of the refrigerant piping. The sensing bulb measures temperature at the evaporator's outlet and sends a signal to the movable rod inside the expansion valve to adjust flow of refrigerant.

95. **Answer A is correct.**
    A stuck-open heater control valve allows the heated air to enter the cooled air side increasing the temperature at the outlet air duct. If the valve were to be stuck closed, it would restrict coolant flow to the heater core and no extra heat would be present. Debris in the evaporator core inlet would not raise the temperature of the air.

Block H. Structural Components, Accessories and Attachments (5 Questions)

96. **Answer C is correct.**
    Mounting power take-offs with insufficient backlash can often produce a whining noise while those running with excessive backlash will produce a clattering sound. One thing worth mention is that the gear design plays a role in the noise factor. Spur gears produce slightly more noise than helical cut gears in operation. Other long term symptoms of insufficient backlash are cracked mounting flanges or drive gear damage.

97. **Answer A is correct.**
    The winch motor is bypassing internally, reducing its ability to perform the task. Control valve spool adjustment would not reduce the capacity of the winch, only slow down its speed if the control valve was not opening wide enough to permit full flow. A restricted case drain would not cause a problem with normal case drain levels. If the main relief valve was set too high, it would not affect the winches speed and capacity in a negative way.

98. **Answer B is correct.**

   The lift cylinder ball and socket joints can be easily tightened by removing shims. Adding shims would loosen it further, and changing out the ball joint may not be necessary until no further adjustments can be made by removing shims. Using a thicker viscosity grease would not help with excessive clearances in the ball joint.

99. **Answer A is correct.**

   The grouser shoe only fits one way correctly with the wide part of the shoe over the link. No shims are required for this procedure. Follow manufactures recommended torque procedures and remove any paint from the contact points as this will allow the grouser shoe to loosen on the track.

100. **Answer A is correct.**

   Either 6 x 19 or 6 x 37 extra improved plow steel (EIPS) with independent wire rope core is a common replacement rope for winches. The improved plow steel (IPS), offers approximately a 15% increase in breaking strength.

Notes

Notes

Notes

Notes

Notes

Notes

Notes

Notes

Notes

Notes